Convoy Ambush Case Studies

Volume I - Korea and Vietnam

Richard E. Killblane
Transportation Corps Historian

US Army Transportation School
Fort Lee, Virginia
2014

Library of Congress Catalog-in-Publication Data

Killblane, Richard E., 1955 -
    Convoy Ambush Case Studies, Volume I: Korea and Vietnam.
/ Richard E. Killblane
    p. cm.
1. Vietnam War. 2. Korean War – 1950. 3. United States Army –
History – 20th century. I.
Killblane, Richard E., 1955- II. Title.

DS

Editing: Gary D. Null
Map Illustration: Stephen A. Campbell
Book Design/Production: Enterprise Multimedia Center

Second Edition

The US Army Transportation School publications cover a variety
of military history topics. The views expressed in this publication
are those of the author and not necessarily those of the
Department of the Army or the Department of the Defense.

ISBN: 978-0-9910108-0-6

Front cover photo: 2½ - ton gun truck Highland Raiders
and gun jeep from the 64th Transportation Company.
Photo courtesy of the US Army Transportation Museum

Back cover photo: Rear view of 2½ - ton gun truck
Assassins from the 541st Transportation Company.
Photo courtesy of the US Army Transportation Museum

Title page photo: Convoy ambush of the 541st Transportation Company
on QL19 near Bridge 5 on 20 June 1970. Assassins
was the last of three gun trucks in the kill zone defending the
disabled tractor and trailer. The road was littered with empty
ammunition cans discarded from gun trucks. Photo courtesy of Larry Wolke

Contents page photo: Ambush of the 359th
Transportation Company on 15 January 1968.
Photo courtesy of Billy Rumbo

Contents

Introduction	4
Korean War	6
Convoy Ambush at Sudong	12
Vietnam War	16
Northern II Corps Tactical Zone	16
2 September 1967	20
11 and 24 November 1967	30
Pre-Tet Offensive	34
4 December 1967	35
15 January 1967	38
Tet Offensive and After	47
31 January 1968 - 14 August 1968	48
88th Transportation Company,	**58**
128th Transportation Battalion	
23 August 1968 - 03 January 1969	58
1969 Doctrine Change	61
523rd Transportation Company, Eve of Destruction	62
22 January 1969 - 22 February 1970	65
New NCOIC	74
Hairpin	76
1 April 1970 - 15 June 1970	77
359th POL	88
9 June 1969 - 16 December 1970	88
Dahl Ambush, 23 February 1971	94
Recovery Mission, 16 December 1970	104
2. Southern II Corps Tactical Zone	**107**
1968 - 1970	111
3. I Corps Tactical Zone	**117**
12 April 1968	118
Lam Son 719, 523rd Transportation Company	**121**
8 February 1971 - 12 March 1971	123
4. III Corps Tactical Zone	**136**
25 August 1968	136
Conclusion	**143**
Appendices: Tactical SOPs	**147**
Bibliography	**149**
Index	**152**

Introduction

When the enemy adopts a policy to attack convoys, truck drivers become front line combat troops. An ambush is an attack with the element of surprise and surprise provides the ambusher an uninterrupted shot and kill. The lethality of the opening volley depends upon the planning and employment of the weapon systems. Yet, no matter how well planned, something will generally survive the initiation of the ambush. After that first shot or volley, the element of surprise is lost and the convoy's reaction creates chaos in the kill zone. It is in this chaos that the truck drivers can turn the fight back on the ambushers. To understand this, convoy commanders must then become tacticians.

Tactics is not something a student of war can expect to learn by reading a manual. To follow a procedure or repeat a technique is to establish a pattern of predictability. There is no one right answer to every question. Each problem requires its own solution. Certain principles, however, remain consistent throughout each problem. The student of war must understand the difference. This concept of war is so vague and elusive that a great number of military philosophers have tried to articulate it into a concept that students can understand.

Because it varies from situation to situation, tactics is not a doctrine. War is chaos. Simply put, in combat each side makes mistakes. The side that protects its weaknesses and exploits that of the enemy, wins. For the infantry, tactics is not a study of battlefield formations and maneuvers but doing whatever is necessary to bring all one's weapons to bear against a weak spot in the enemy position and exploiting it. It should not be much different with convoys. Most victories are determined at one decisive point in the battle. The trick is finding that location that provides the tactical advantage at the decisive time. Winners train to make this a habit.

Mission, enemy, (friendly) troops, and (weather and) terrain (METT) is probably the simplest way to understand tactics. While the mission of the infantry is to close with and destroy the enemy, the mission of the transportation corps is to deliver the cargo. The destruction of the enemy facilitates this mission but does not become the mission. A tactician has to think like a hunter. A successful hunter thinks like his prey. A tactician has to think like his enemy. Only by understanding how his enemy thinks can the tactician predict his enemy's next moves. By anticipating what the enemy will do next, and then the tactician can plan to exploit the vulnerabilities of his enemy. In a convoy ambush, the convoy is the prey and the ambusher the hunter. Through thorough knowledge of his own troops, the tactician can defend his weaknesses and apply his strengths to the enemy weaknesses. The tactician must constantly be aware of the terrain and how it provides an advantage to the enemy and how he might use it to his advantage.

However, there are too many uncertainties in combat. A tactician must be flexible just as a fighter. A fighter trains his body through repetition to instinctively move in a certain way with maximum ease and power. These would be the equivalent of battle drills. He studies his opponent. He looks for patterns so he might predict where his opponent might leave himself open. He also looks at the opponent's strength, which he then plans to nullify or avoid. The fighter studies his opponent's strengths and weaknesses then plans to protect his weakness and match his own strength against his opponent's weakness. If it was that simple the fight would be over quickly. Combat is not that simple. The opponent has also done his own study.

In an academic discussion with a martial artist, they can point out every counter that would defeat any technique one might use, but one does not defeat a martial artist in a discussion though. It is done in a fight. A boxer has only three punches, the straight, the hook or cross and the upper cut. With just these three punches, they win fights. The mixed martial arts employs a multitude of punches and kicks each designed to strike a specific weak spot on the body, but in a fairly even contest, the basics usually determine the outcome. Timing is what separates the winners from the losers. The fighter wins by instinctive feel. He throws out punches and kicks in combinations until in a flash he sees an opening and instinctively strikes at it with his strongest punch. The time that elapses between thought and action is almost instantaneous because of training, what is recognized as muscle memory. Connecting with the right strike and force determines a knockout.

To master the art of war, the tactician must train his mind and body just as a fighter. The tactician trains his mind through an academic study. For a warrior leader, the commander represents the head and the organization represents the body. He should train his organization as a fighter trains his body. Battle drills or immediate reaction drills represents the building blocks of tactics. Like a fighter trains to perfect the punch or kick, the professional warrior trains his organization to perform its drill with the same level of perfection instantaneously. With the battle drills in place, the tactician then spars with an opponent in war games to bring the mind and body together.

How to study war? The student of tactics studies previous fights and mentally places himself in the position of the participants. Knowing what they knew, how would he have reacted instead? In hindsight, what was the best course of action, remembering that there is no one perfect solution? Any number of actions would have succeeded. The tactician must learn what would have worked best for him.

For this reason, I have pulled together relevant examples of convoy ambushes. The 19th century, the Vietnam War and recent wars in Iraq and Afghanistan provide a wealth of examples of convoy ambushes from which to study. Unfortunately, the US Army did not record many good accounts of ambushes during the Vietnam War. Much of what is presented in this text is based upon oral interviews of the participants, sometimes backed by official record, citations or reports. For this reason, some of the ambush case studies present only the perspective of a crew member of a gun truck or the convoy commander. Since this academic study works best when one mentally takes the place of one of the participants, this view of the ambush serves a useful purpose. After my own review of the ambushes, I have added the lessons of the participants as well as drawn my own conclusion as to what principles apply to convoy ambushes.

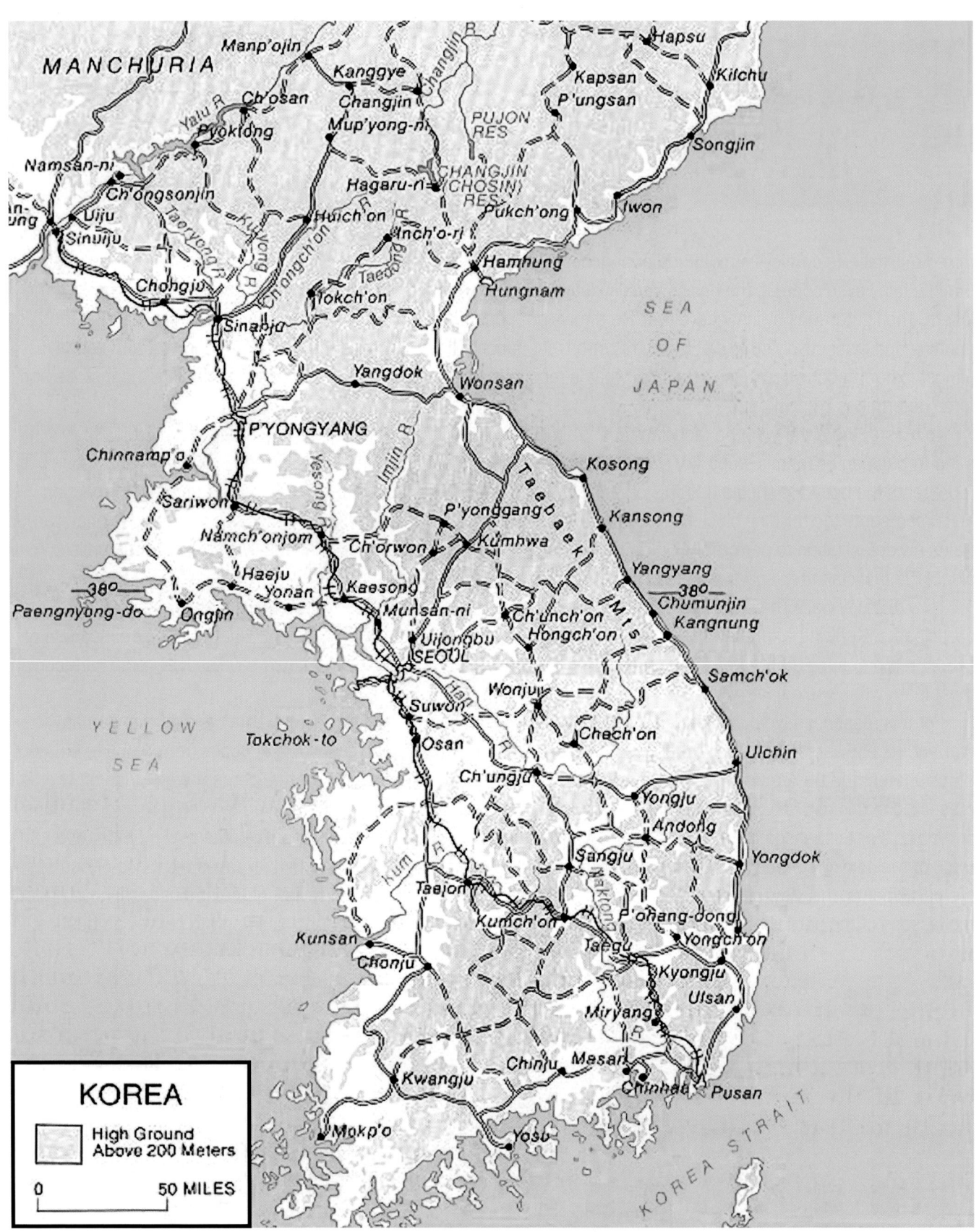

Ebb And Flow, November 1950 - July 1951
Center of Military History, 1990
Map courtesy of Billy C. Mossman

Korean War

Convoy Ambush at Sudong
10 December 1950, 52nd Transportation Battalion

During its short history, the Transportation Corps claims three recipients of the Medal of Honor all for actions in convoy ambushes. When the lines between the enemy and friendly forces become fluid or non-existent, great acts of heroism are required of Transporters. In October 1950, 47-year old, Lieutenant Colonel John Upshur Dennis Page was assigned to the X Corps Artillery but his actions in the role of a Transporter would earn him the Medal of Honor. He had spent World War II training artillerymen at Fort Sill, Oklahoma and saw no service overseas but was eager to finally experience combat in Korea. On 27 November, Page arrived with the X Corps at Hamhung on the east coast of the Korean Peninsula. The 1st Marine and 7th Infantry Divisions pushed north to the Chinese border.[1]

On 29 November 1950, the Chinese Communist Army (CCA) conducted a large scale infiltration attack across the Chinese border to cut off and destroy the 1st Marine Division and the 7th Infantry Division defending around the Changjin (Chosin) Reservoir, Korea. The 377th Transportation Company had been attached to the 7th Infantry Division during this operation. The 3rd Infantry Division was the X Corps reserve and defended the main supply route from below Chinhung-ni to the port of Hungnam. The Army's Task Force Faith, organized around the 31st Infantry, fought desperately to link up with the 1st Marine Division at Hagaru-ri. The Marines absorbed what remained of the battered Task Force Faith and fought their way back to a near completed airstrip at Hagaru-ri. While the Marines evacuated their dead and wounded, the Chinese infiltrated past the force and established roadblocks along the route between it and Koto-ri. MG Oliver P. Smith, Commander of the 1st Marine Division, needed to keep the main supply route (MSR) from Hagaru-ri to the base of the mountain pass at Chinhung-ni secure, so he positioned a battalion of the 1st Marine Regiment at Koto-ri at the top of the mountain and another at Chinhung-ni at the bottom.[2]

The enemy tactics was to infiltrate behind the Americans and establish road blocks along the narrow mountain roads with fallen trees, piles of rocks or craters especially around curves. They would then wait until the lead elements of the retreating column had entered the kill zone then attack the column with mortars, automatic fire and small arms

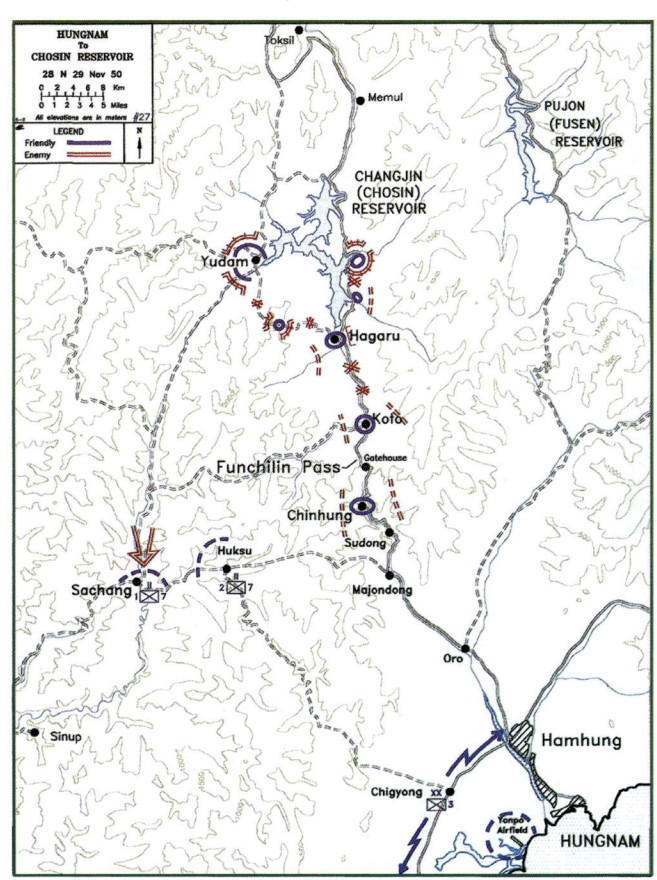

Hungnam to Chosin Reservoir Map.
Produced by the Center of Military History

fire from positions concealed in the ridges and defiles. The Chinese preferred to wait until darkness fell at 1745 hours.

The 5th and 7th Marine Regiments and the remnants of Task Force Faith arrived at Hagaru-ri to begin the air evacuation of their wounded. On 29 November, LTC Page left X Corps headquarters and was attached to the 52nd Transportation Truck Battalion with the mission to establish traffic control on the main supply route to Marine positions and those of some Army elements on the Chosin Reservoir plateau. For the next 11 days, Page would serve in the role of a Transportation Corps officer.

LTC Page and his jeep driver, CPL David E. Klepsig, fought their way past a Chinese machine gun position and reached Koto-ri. Having completed his initial mission, Page was free to return to the security of Hamhung but chose to remain on the plateau to aid an isolated signal station, thus cutting himself off with elements of the Marine division. After rescuing his jeep driver by breaking up an ambush near a destroyed bridge Page reached the lines of a surrounded Marine garrison at Koto-ri. He then voluntarily developed and trained a reserve force of assorted Army troops encircled with the Marines. By exemplary leadership and tireless devotion, he made an effective fighting unit available. In order that casualties might be evacuated, an airstrip was improvised on frozen ground partly outside the Koto-ri defense perimeter which was constantly under enemy attack. During two such attacks, Page exposed himself on the airstrip to direct fire on the enemy, and twice mounted the rear deck of a tank, manning the machine gun on the turret to drive the enemy back into a no man's land.[3]

On 3 December while being flown low over enemy lines in a light plane, LTC Page dropped hand grenades on Chinese positions and sprayed foxholes with automatic weapons fire from his carbine. On 7 December, the lead elements of 7th Marines arrived at the airfield at Koto-ri. There the rest of the Marines and Soldiers would gather for evacuation more than 500 wounded accrued along that retreat and to assemble for the final retreat ten miles of treacherous winding road down the mountain.[4] Once the Army and Marines arrived at Koto-ri, LTC Page flew to Hamhung to arrange for artillery support of the beleaguered troops attempting to break out. Again Page refused an opportunity to remain in safety and returned to assist his comrades.[5]

MG Smith needed a mobile force organized around a battalion of infantry to clear the ten miles between Koto-ri and Chinghung-ni.

LTC John U.D. Page by his jeep in Korea.
Photo courtesy of Margaret S. W. Drew and American Battle Monuments Comission

He asked LTG Edward M. Almond, X Corps Commander, to provide a relief force for his Chinghung-ni force. Almond tasked the 3rd Infantry Division with the mission. On 6 December, the 3rd Infantry Division formed Task Force Dog around the 3rd Battalion, 7th Infantry Regiment. The 52nd Transportation Truck Battalion, commanded by LTC Walden C. Winston, moved Task Force Dog into Chinhung-ni to replace the 1st Battalion, 1st Marines on 7 December, so the latter could push up the mountain and secure the key terrain along the pass. The other two battalions of the 65th Infantry Regiment took positions to protect the road south of it, but did not have enough men to occupy all the key terrain and adequately defend the route. So on 8 December, the 65th Infantry sent G Company to Sudong where one platoon occupied the north side of the road and the rest of the company occupied the hills west of the road about a mile south of Sudong. Likewise, the narrow roads and steep ridge made for poor radio communications between the two groups.[6]

On 9 December, the Marines secured the high ground along the route and the 1st Engineer Battalion had erected a Treadway Bridge over the blown bridge one-third of the way down the mountain. Smith planned to push the trains down first and then have the Marines leave the high ground and follow after the trains passed. MG Smith assigned the tank battalion to the rear of the column; because if an M-26 tank stalled or threw a tread along the one-lane mountain road, the rest of the vehicles could not drive around it. The 52nd Transportation Battalion arrived at Koto-ri on 9 December and the next day, the column of 1,400 vehicles from the 7th Marine Motor Transport Battalion, the Division Trains and the 52nd Transportation Battalion started down the mountain from Koto-ri.[7]

The road snaked through a valley and Sudong was a small village of houses that flanked the road. By the time the Army trucks approached Sudong, temperatures had dropped to below zero. The trucks were loaded to their full capacity with Marines and dead Marines were tied to the bumpers and fenders. Division Trains Number 1 and 2 cleared Chinhung-ni and reached Hamhung that afternoon. The empty trucks then turned around to pick up troops.[8]

LTC Page had joined the rear guard and when the column neared the entrance to a narrow pass, it came under frequent attack on both flanks. As feared, a tank jammed into the bank blocking the road. Unable

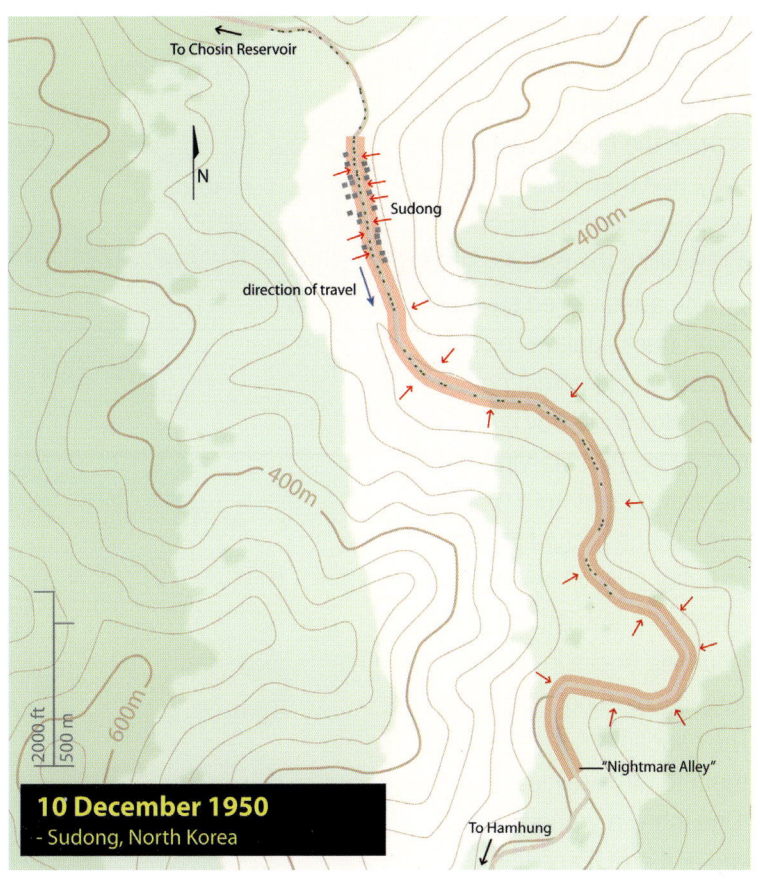

10 December 1950 - Sudong, North Korea

to extract it under fire, the tank crews disabled the six tanks behind it and proceeded down the mountain on foot. Meanwhile Page mounted an abandoned tank and provided heavy machine gun fire for passing vehicles until the danger diminished. Later when his section of the convoy was threatened in the middle of the pass, Page took a machine gun to the hillside and delivered effective suppressive fire while men and vehicles passed through the ambush.[9]

A force of Chinese from the 89th Division had infiltrated as far south as Sudong just south of Chinhung-ni. Fifty or more Chinese infiltrated the village of Sudong and set up ambush positions west of the road and neither element of G Company, 65th Infantry knew this. Just after dark on the night of 10 December, an estimated Chinese force of 200 attacked G Company positions south of Sudong but were repulsed with heavy losses. An approaching Marine column halted while the fight took place and upon the withdrawal of the Chinese, the Marine column continued unmolested. After the news of the attack reached Task Force Dog headquarters at Chinhung-ni, they halted all road traffic south of there. Shortly after dark the Marine Liaison Officer to Task Force Dog received word the road below Sudong was clear and reopened it for traffic. Nightlong 155mm artillery fire on the enemy positions on the west side of the road missed their target by 250-300 yards due to a meteorological error. So the barrage had no effect on the enemy.[10]

The 1st Marine Regimental Train reached the bottom of the pass just after it turned dark and as it entered

Korean War 9

Sudong, Chinese soldiers rushed from around the huts firing their submachine guns and throwing grenades at the convoy. The lead M-39 armored utility vehicle of the anti-tank section rammed a truck the enemy had placed in the road to block the convoy, and then cleared the kill zone. A Chinese soldier tossed a grenade in the open compartment of another M-39 killing the passenger and severely wounding the driver. The attack killed several drivers and set at least five lead trucks ablaze thus halting the column. If the retreating column did not start moving soon, the enemy would slowly destroy it.[11]

LTC Page, his Marine jeep driver, PFC Marvin L. Wasson, and another Marine dismounted from their jeep and then ran up the road to the scene of the fighting. The Marine paused to fire at the enemy, while Page and Wasson ran past burning vehicles, stumbling over fallen Marines. Illuminated by the burning trucks in the kill zone, Page and Wasson charged into about 30 enemy soldiers firing as they ran. Their audacity surprised the Chinese, causing many of them to run away leaving several of their comrades. One threw a grenade which wounded Wasson in the head and arm. Page told Wasson to turn back and he would provide covering fire. Wasson staggered back and turned just in time to see Page charge after the fleeing Chinese. Page did not return, but his one-man attack bought enough time for LTC Winston, Commander of the 52nd Transportation Battalion, to organize the Marines and Army truck drivers for a counter-attack.[12]

Wasson, bandaged by then, returned to the fight firing his 75mm recoilless rifle while a machine gun covered him. Wasson fired two white phosphorus rounds at a hut, which he felt was the enemy strong point. The hut erupted into flames while Chinese soldiers ran to escape a fiery death only to be met by machine gun fire. Wasson then helped push the burning trucks with ammunition off the road to clear a path for the convoy. The fight lasted for several hours and LTC Winston's counter-attack drove off the last remnants of enemy resistance as the sun came up. The Chinese ambush at Sudong killed eight Americans and wounded 81, destroyed nine Marine trucks and one M-39 armored utility vehicle. As the convoy continued through Sudong, it came upon the body of Page in the road with 16 dead Chinese near him. When Task Force Dog

LTC John U. D. Page during World War II.

passed through the village later that afternoon, they saw five trucks still smoldering beside the road and four American dead, three badly burnt that had not yet been carried away.[13]

The US Marine Corps initially rewarded LTC John U. D. Page posthumously with the Navy Cross, but when his full story became known in 1957, an act of Congress upgraded his award to the Medal of Honor. Although a Field Artillery officer, in his first time in combat, LTC Page served in the role of a Transportation Corps officer coordinating the withdrawal of the 1st Marine and 7th Infantry Divisions. Warriors do what they know is right without regard to the consequences. LTC Page always did what he saw was needed and never backed away from danger.[14]

Lesson

Using the cover of darkness, the Chinese ambushed a convoy of Marine trucks passing through Sudong and successfully stopped the convoy by destroying between five and nine trucks. Without ring mounts and machine guns, the standard reaction to a convoy ambush was to dismount and fight through like infantry. During the ambush at Sudong, LTC Page led two men in a quick counter-attack that bought time for LTC Winston to lead truck drivers in attacks that successfully beat back the enemy until the sun came up and they could clear the road of disabled vehicles. Dismounting and fighting through the kill zone is an effective reaction if no other fire power is available.

Once the Korean War settled into static trench warfare, the Chinese continued to infiltrate behind UN lines and ambush convoys. The convoys then added ring mounts to every third truck for self protection and initially mounted .30 caliber machine guns on them, but found the M-2 .50 caliber machine gun intimidated the enemy more and became the preferred weapon. So a gun truck became any truck with a crew-served weapon mounted on it. Once armed in this manner, truck drivers did not have to dismount to defend the convoy, but could return fire and clear the kill zone. The US Marine Corps, even into Operation Iraqi Freedom, maintained the policy that all Marines were riflemen first and therefore drivers would dismount and attack through the enemy in the event of a convoy ambush. This happened to be the same policy when US Army truck units arrived in Vietnam.

If an ambush is planned well, the kill zone is not a place to remain and the advantage trucks have over dismounted infantry is mobility and speed. Since it was important to clear the kill zone, the dismounted infantry doctrine called for them to turn in the direction of fire and fight through the enemy; but if the road is not blocked, trucks can clear the kill zone faster by driving out of it.

The Korean War set the weapon to task vehicle ratio at one-to-three. A gun truck was just another task vehicle armed with a machine gun, not a dedicated fighting platform, so it needed to clear the kill zone with the rest of the vehicles hauling cargo. However, if trapped in the kill zone, the truck drivers and in the case of any infantry passengers would have to dismount to fight. Not until the Vietnam War was a gun truck designed as a dedicated weapons platform.

---

[1] LTC Roy E. Appleman, *Escaping the Trap; The US Army X Corps in Northeast Korea, 1950,* College Station, Texas: Texas A&M University Press, 1990.
[2] Letter from MG Oliver P. Smith, Commanding General, 1st Marine Division, to the Commandant of the Marine Corps, 17 December 1950.
[3] LTC John U. D. Page Medal of Honor Narration.
[4] Smith letter.
[5] Page Medal of Honor Narration.
[6] Appleman, *Escaping the Trap;* Smith letter; and Billy C. Mossman, *Ebb and Flow; November 1950-July 1951,* Washington, DC: Center of Military History, 1990.
[7] Smith letter.
[8] Smith letter.
[10] Appleman, *Escaping the Trap.*
[11] George E. Petro, Appendix 10 to Annex Peter Peter to 1st Marine Division Special Action Report, 30 December 1950.
[12] Appleman, *Escaping the Trap.*
[13] Page MOH Narration; Appleman, *Escaping the Trap;* "Springing the Chosin Reservoir Trap," L. B. Puller, Annex Peter Peter to 1st Marine Division Special Action Report, Headquarters, 1st Marines (Reinf), 15 January 1951; and LTC Roy E. Appleman Papers, US Army Historical and Education Center.
[14] Appleman, *Escaping the Trap.*

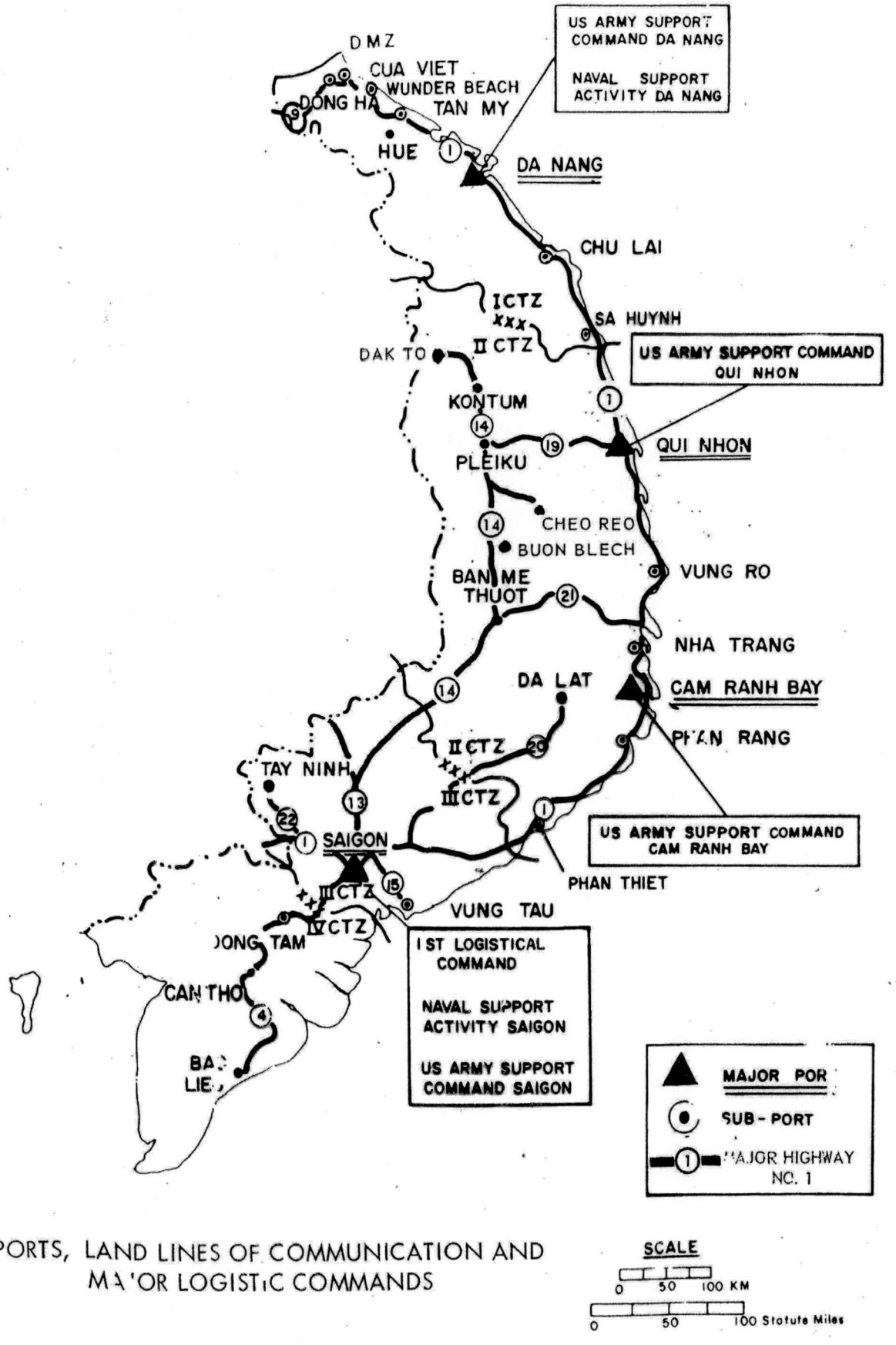

PORTS, LAND LINES OF COMMUNICATION AND MAJOR LOGISTIC COMMANDS

Vietnam War

The US Army assumed a greater role in ground combat in South Vietnam in the summer of 1965, and there would be three incremental buildups of forces over the next three years. For command and control, Military Assistance Command, Vietnam (MACV) divided the country into four Corps Tactical Zone with I Corps in the north, II Corps in the Central Highlands, III Corps encompassing the plains around Saigon, and IV Corps in the Mekong Delta. To reduce the ground line of communication and vulnerability to convoy ambush, the logisticians chose sub ports versus a central hub and spoke operation originating out of the major commercial port at Saigon. The US Navy established beach ramps at Hue-Phu Bai and Dong Ha in I Corps. The 1st Logistics Command established two sub ports, Qui Nhon in Northern II Corps near the intersection of the coastal highway Highway 1 and Route or Quoc Lo (QL) 19 leading to Pleiku, and Cam Ranh Bay in Southern II Corps. The 1st Logistics Command also built Newport on the Saigon River just upriver from Saigon as the hub for logistics in III Corps. From the Saigon River into IV Corps, river and canal provided the best means of hauling cargo. The threat and solution to convoy ambush was different in each area.

The enemy was either the homegrown Viet Cong (VC) who lived in the area he fought. He was the farmer by day and guerrilla by night and usually not as well trained as the North Vietnamese Army (NVA) Regular sent down from the North. The VC tended to operate in small guerrilla cells about squad size and more often conducted small ambushes. The more disciplined NVA operated in larger formations and had the time to plan and rehearse large scale combat operations. The Vietnamese had honed their guerrilla tactics fighting against the Japanese invaders in 1941-1945, and then as the Viet Minh against the French during the Indochina War, 1946-1954. The new version of an old enemy sparred with the American's air cavalry concept over the next two years only to learn the combat forces relied on truck convoys to get everything they needed to fight.

Coordinating the fires of men in a kill zone provided the greatest challenge so the ambushing forces rarely exceeded 50 men in a given kill zone. The NVA organized their ambushing force into security, support and assault; and employed a number of types of kill zones. The simplest kill zone was the linear kill zone which could hit a convoy heading either direction along the road. They would just fire at anything in front of them. An L-shaped kill zone usually placed a crew served weapon on the other side of the road or at the bend in the road. With support elements on the other side of the road allowed the ambusher to catch the convoy in the cross fire, but required greater fire control measures to prevent the ambushers from firing into each other. A limitation of the L-shaped kill zone was it was most effective catching a convoy heading in one direction – toward the base of the "L." Z-shaped kill zone was merely two L-shaped kill zones linked together. They could catch a convoy coming either way. The V-shaped kill zone was most effective in a mountain pass, like Ban Me Thout in Southern II Corps. The support force with the crew served weapons was usually placed across the valley on opposite ridge and the assault force above on the road on the other side. The two ridges would link into a "V" while the two elements of the ambush appeared more parallel to each other. The enemy also learned to set up multiple kill zones, in what was called an area ambush. As compared with later wars, the NVA ambush tended to be more violent than those experienced by the US Army in Iraq or Afghanistan.

14 Convoy Ambush Case Studies - Volume I

Top left: M-35 2½-ton Cargo Truck. *Left Center:* M-52 5-ton Tractor Refrigeration Van "Reefer." *Left bottom:* M-54 5-ton Cargo Truck. *Top:* M-52 5-ton Tractor, M-126 Trailer, Stake and Platform (S&P). *Center:* M-52 5-ton Tractor, M131 5,000-Gallon Fuel Tank Trailer. *Bottom:* The gun jeep of the 8th Group is the M-151, ¼-ton truck. A normal crew was a driver, and vehicle commander who also manned the machine gun. Photos courtesy of the US Army Transportation Museum

Vietnam War 15

1. Northern II Corps Tactical Zone
8th Transportation Group

The 8th Transportation Group had three truck battalions that hauled cargo back and forth along Route, or Quoc Lo (QL), 19 through the Central Highlands. The 27th Transportation Battalion, which arrived in the summer of 1965, consisted of primarily medium trucks, M52 tractors pulling M126 trailers. The 54th Transportation Battalion, which arrived the next summer, had the light trucks, M54 5-ton and M35 2½-ton cargo trucks. Both of these battalions were garrisoned in the vicinity of Qui Nhon and marshaled at Cha Rang Valley every morning for the long haul to Pleiku, 110 miles to the west. The 124th Transportation Battalion, which arrived during the summer of 1967, had both light and medium truck companies which picked up cargo at Qui Nhon or pushed it out to the camps along the Cambodian border. Generally, two convoy serials launched out of the marshalling yard near Qui Nhon every morning, one with light trucks (2½-ton and 5-ton cargo) under the control of the 54th Battalion, and another of Stake and Platform (S&P) (5-ton tractors and trailers) under the control of the 27th Battalion. The 124th Battalion ran mixed convoys back to Qui Nhon in the morning or light trucks out to the camps. The westbound convoys along QL19 were called "Friscos," and the eastbound convoys out of Pleiku were called "New Yorkers."

By September 1967, QL19 was a two-lane unimproved road that ran about 35 miles along the coastal plain then snaked up a mountain to the An Khe Pass. At one point below this pass, the road switched back on itself in a sharp turn the drivers called "The Devil's Hairpin" or "Dead Man's Curve" for short. Traffic generally slowed to a crawl of 4 miles per hour at that turn regardless of whether the convoys were heading up hill or down. Once over the pass, the road leveled out but pot holes as deep as a foot kept traffic to 15 miles per hour. Right before Pleiku, the road again rose up to meet Mang Giang Pass then leveled out onto the Highland Plateau where their destination awaited.

For tracking the progress of the convoys, they reported the bridge numbers as check points. There were 36 bridges along the 110 miles of QL19 and tanks and APCs provided convoy security for all but bridges 1, 6, and 9 along the coastal plain. Check points were referred to by their nearest mile marker. The Korean Tiger Division had responsibility for the first eight check points at the bridges from Qui Nhon to the base of mountain. All the Korean soldiers had lived through the Korean War, 1950-53, and hated communists. Some of the field grade officers even served in the Japanese army during WWII. They welcomed the opportunity to fight them anywhere. They were extremely professional and were proud to serve in Vietnam. Their method of

This obelisk marks the location on Mang Giang Pass on QL19 where Groupemente Mobile 100 was ambushed leading to the complete and annihilation of this French Brigade in 1954. Photo courtesy of Bill Eichenberg

responding to enemy resistance was brutal and often times involved civilians in the villages where the attacks occurred. Consequently, the enemy did not launch many attacks in the Republic of Korea (ROK) sector and the drivers felt safe. However, the Koreans did not initially guard the slope leading up to An Khe Pass.

Up until September 1967, there were no convoy ambushes other than occasional sniping or planting of mines in the road. For that reason, convoy commanders did not concern themselves with the interval between vehicles in the mountains. More often, the faster trucks pulled up bumper-to-bumper and pushed the slower trucks in front of them when going uphill. It was not uncommon to see a number of trucks driving in tandem. The 8th Group usually kicked out two convoy serials of 30 to 40 vehicles early in the morning. Convoys were generally grouped by type of trucks; light truck convoys fell under the control of a 54th Battalion convoy commander and the medium trucks fell under the 27th. Within each serial, trucks were generally arranged by load; the heavier or more explosive loads such as fuel and projectiles were in the rear.

Convoys kicked out early in the morning usually reaching Pleiku by noon. After either unloading their cargo or switching trailers, they returned that afternoon reaching Cha Rang Valley by dark. Convoys did not run along QL19 at night, so there was only one run made per day.

The battalions had standard operation procedures (SOP) for reaction to ambushes but since they had not encountered any, most Soldiers did not even know their SOP. The 54th Battalion had been in Vietnam since October the year before. Major Nicholas Collins, 54th Battalion S-3, had consulted with other truck and infantry units as to an appropriate reaction to any particular threat. The popular consensus that ended up in the SOP read, "If caught in an ambush, halt in the center of road (shoulders may be mined). Take cover and return fire in the direction of the enemy, and be prepared to assault the enemy position and to fight your way out." This doctrine had not changed since the Korean War. At that time, the drivers of 8th Group were armed with the M-14 rifle instead of the shorter M-16. Since the doctrine was to dismount the vehicle and return fire, the length of the weapon was immaterial. The convoy commander and assistant convoy commanders usually rode in jeeps mounted with M-60 machine guns. That would have been the only armed escort.

One of the companies of the 54th, the 666th Light Truck, had recently arrived that August. It had been assigned to Fort Benning, Georgia, where it supported the US Army Ranger School among other duties. The drivers were probably the only drivers in the 8th Group who had any reaction to ambush training.

5-ton cargo trucks of the 541st Transportation Company taking a bypass around a bridge. The soldiers standing behind the cab with M16s were pulling security. Photo courtesy of the US Army Transportation Museum

Vietnam War 17

Ambush Alley - Portion of road between An Khe and Mang Giang Pass. In the 2 miles represented in this aerial photo there were 6 ambushes on the 8th Group Convoys in 11 months since the first complex ambush in September 1967. Note the area adjacent to the road which has been cleared by rome plow.

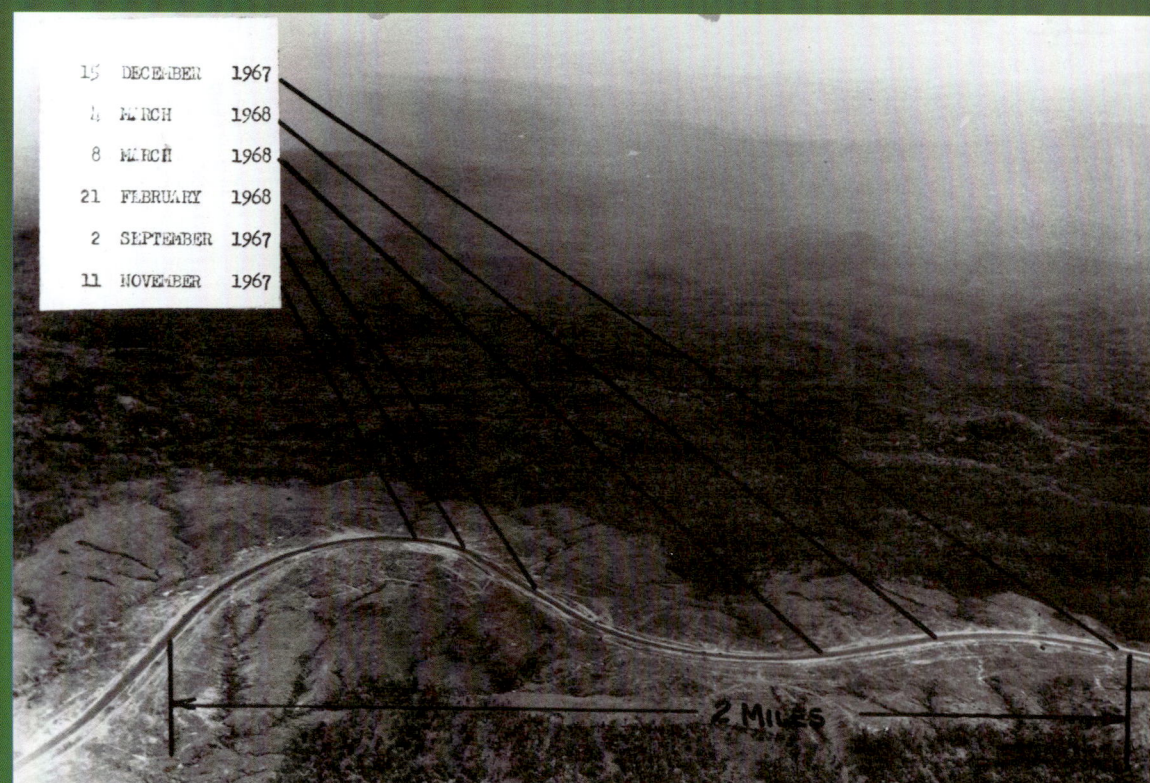

18   Convoy Ambush Case Studies - Volume I

13 FEBRUARY 1968
14 AUGUST 1968
23 AUGUST 1968
15 MARCH 1968
6 FEBRUARY 1968
31 AUGUST 1968
31 JANUARY 1968
30 JANUARY 1968
8 MARCH 1968
21 JANUARY 1968
10 SEPTEMBER 1968

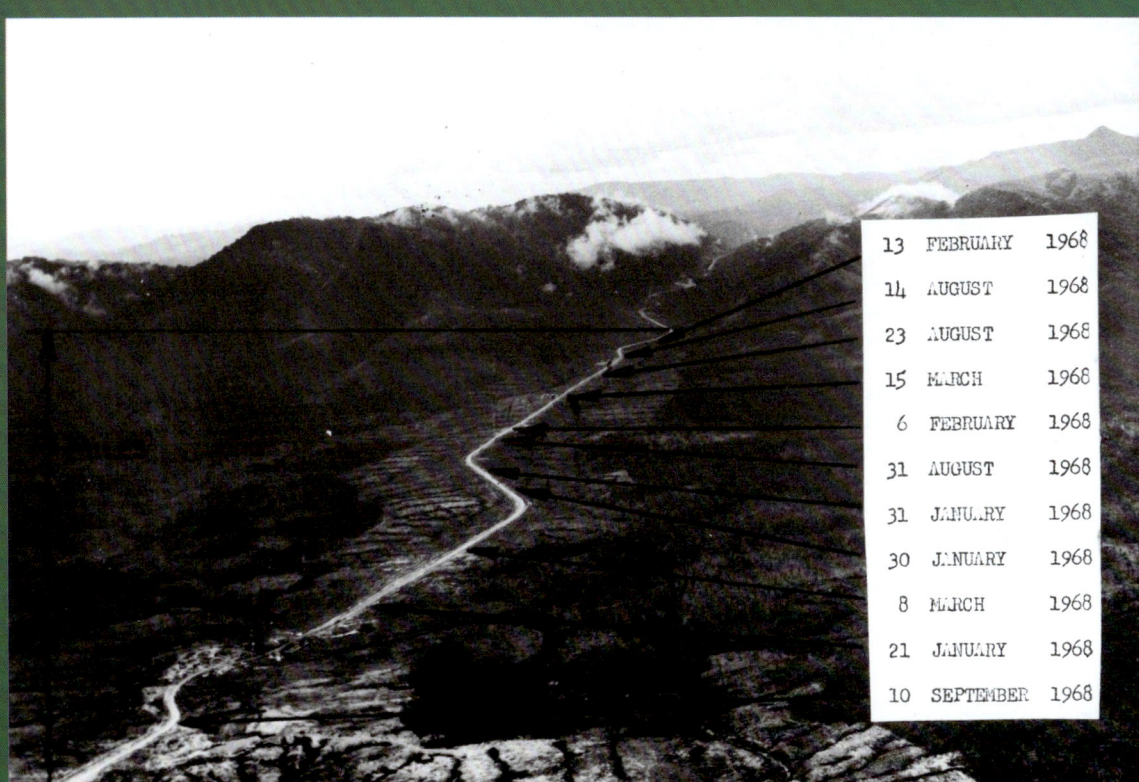

Mang Giang Pass - This 3 mile stretch of road approaching the base of Mang Giang Pass, was the most dangerous portion of road from Qui Nhon to Pleiku. There were 11 ambushes on the 8th Group convoys in the past 11 months since the first complex ambush in September 1967.

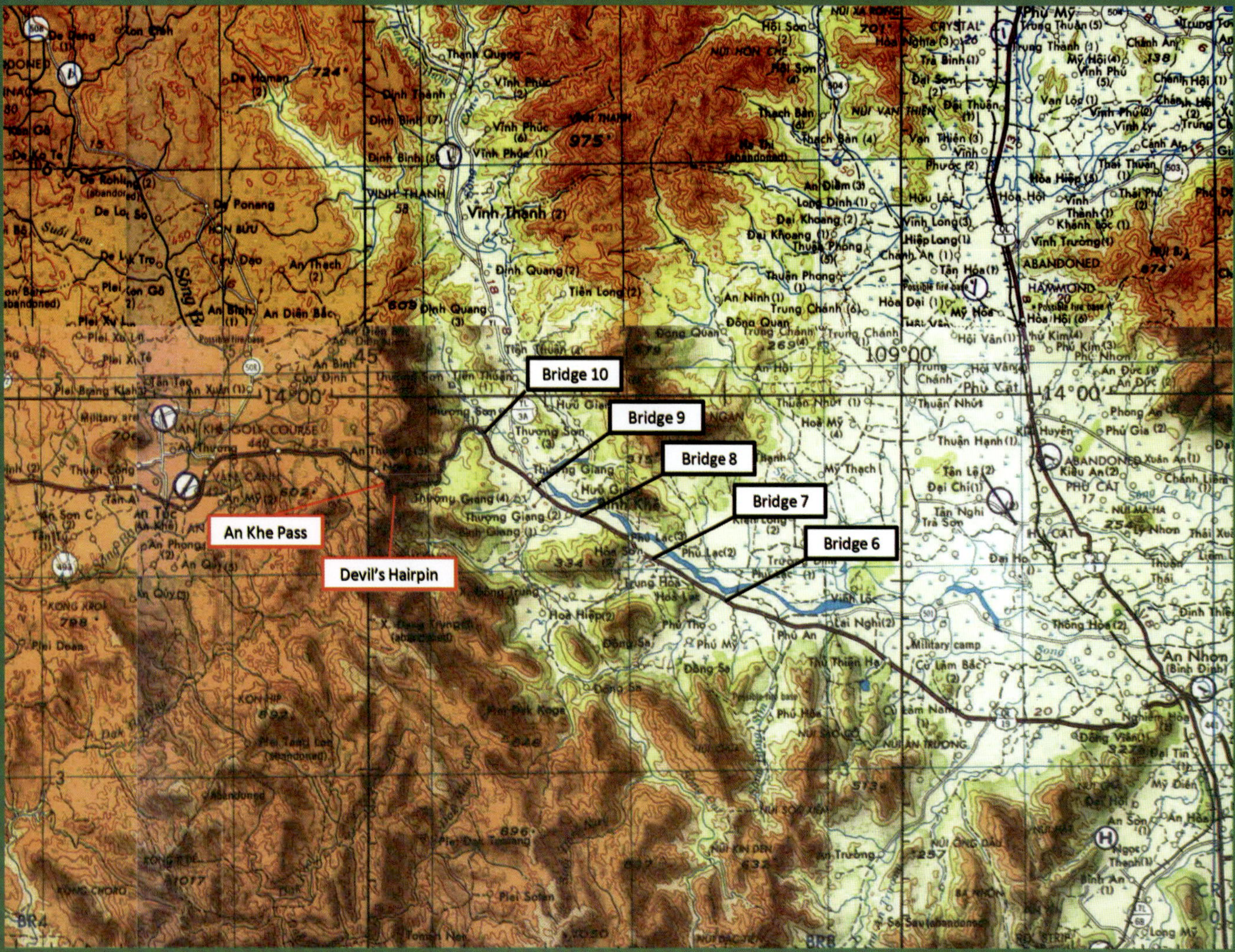
Bridge 10
Bridge 9
Bridge 8
Bridge 7
Bridge 6
An Khe Pass
Devil's Hairpin

The First Large Scale Convoy Ambush 2 September 1967
54th Transportation Battalion

On 2 September, the North Vietnamese Army (NVA) changed their tactics. They found the weakness to the American air assault concept. Realizing that the mechanized infantry and air cavalry at An Khe and Pleiku were entirely dependent upon trucks for supplies, the NVA attacked the supply line.

Convoys marshaled along QL1, the coastal highway near the intersection with QL19 for the run to Pleiku every morning. They lined up by type with the 54th Transportation Battalion in charge of light trucks, M54 5-ton and M-35 2½-ton cargos, and the 27th Battalion in charge of the M-52 5-ton tractors and M126 trailers. The crews consisted of one driver each armed with an M-14 rifle and about four or five 20-round magazines.

A few truck companies had the new M-16 with four or five 20-round magazines. The trucks came from any of the companies in each battalion. The two march units would lead out at separate times and would travel the 110 miles to Pleiku by noon. If they could unload before the MPs closed the road, then they were free to return that day, otherwise they had to remain overnight at Pleiku. The trucks returning from Pleiku generally did not travel in convoys, but just as single vehicles. The road from Qui Nhon to An Khe was not dangerous, but the road from An Khe, especially once Mang Giang Pass was cleared, was dangerous and could not be traveled after nightfall. The truck drivers from Qui Nhon had two choices; drive to Pleiku and stay there over night, or drive to Pleiku and return to Qui Nhon in one day. The one-day trip was the preferred choice but the truck driver had to clear a checkpoint east of Pleiku by a certain time or he could not leave for Qui Nhon. As it got later in the

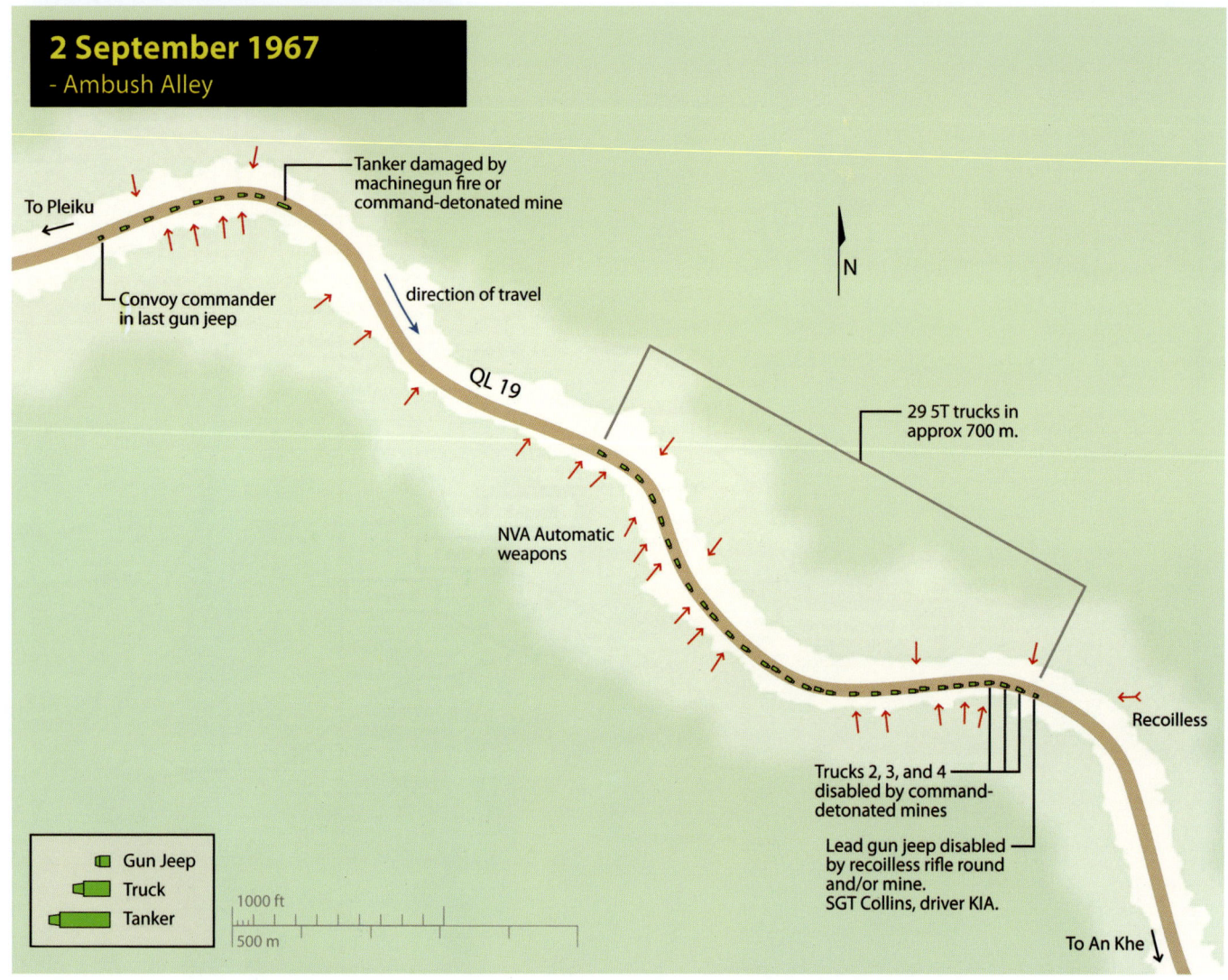

day drivers rushed to unload and take off to clear the checkpoint. This rush produced a high density of trucks late in the day and long line of trucks following each other slowly along QL19.[15]

An eastbound convoy of 90 trucks from both battalions was returning that afternoon from Pleiku under the protection of only two jeeps with M-60 machine guns. CPT Paul Geise, Commander of the 523rd Transportation Company, was the convoy commander riding in the last vehicle in the convoy.[16] SGT Leroy Collins rode in the lead gun jeep. The 54th Battalion had control of lead serial of 37 cargo trucks, which consisted of trucks from its different companies. The convoy had just descended the seven kilometers of Mang Giang Pass and picked up speed across the plateau heading toward An Khe. Because of mechanical problems, a 5,000-gallon tanker created a 500-meter gap between it and the lead 29 vehicles as they approached the winding stretch of road between Check Points 96 and 89. At that time, the jungle grew up to about eight or ten feet from the road.[17]

The standard operating procedure (SOP) for the highway patrol from C Company, 504th MP Battalion was to proceed west from An Khe to Check Point (CP) 102 below Mang Giang Pass to link up with highway patrols from B Company each morning. The highway patrol consisted of two gun jeeps, standard jeeps modified with armor and a post for mounting an M-60 machine gun. As they drove west toward CP 102, the gunners would fire into any locations along the road that might harbor enemy ambushers, a technique known as "recon by fire." Upon arriving at CP 102 and linking up with the B Company highway patrol, C Company would radio back to Qui Nhon that the road was open and trucks could begin their drive to Pleiku. At the end of the day, B Company highway patrol jeeps would wait at the Pleiku checkpoint until the last truck was had passed. The MP jeeps followed the trucks to CP 102 where the highway patrol jeeps of 3rd Platoon, C Company would wait at the end of the day for the last truck and then following them to An Khe. Any trucks that made it past the Pleiku checkpoint were allowed to continue from An Khe to Qui Nhon even after nightfall. 1LT Thomas Briggs' 3rd Platoon highway patrol's day ended when it drove into An Khe behind the last truck.[18]

At 1855 hours that evening, as the convoy snaked around a series of curves, an NVA company struck the lead gun jeep with a 57mm recoilless rifle round killing SGT Leroy Collins and a claymore mine mounted on sticks level with the driver's head detonated on the left front of the vehicle wounding the driver and gunner. Simultaneously, disabled the tanker in the convoy and trapped the lead convoy.[19] The enemy pulled boards across the road with mines on them and detonated them in front of the next three trucks behind Collin's.[20] The enemy also sprung a secondary ambush on the other half of the convoy setting the tanker on fire. An estimated 60 to 80 enemy soldiers were dug in about 30 yards up the hill firing down on the trucks.[21]

J.D. Calhoun, of the 666th, was driving his 2½-ton truck eighth in line of march. He barely heard the firing

A recoiless rifle made a direct hit on the passenger side of the jeep killing SGT Collins and a claymore detonated on the left front. Photos courtesy of Walter Medley

Vietnam War 21

These two trucks show damage by mines pulled across the road. Photos courtesy of Walter Medley

of small arms over the roar of his diesel engine. Calhoun did not realize he was in an ambush until he saw the impact of bullets on the truck ahead of him, which came to a halt. He thought, "Oh crap. I can't sit in a truck. I've got to get out and get behind something." The drivers were taken by surprise. Many did not know what to expect. The kill zone spread out over 700 to 1,000 meters. Later investigation revealed the enemy had dug and camouflaged fighting about 50 positions along the high ground and, from fecal evidence, had occupied them for several days. The drivers did not know the 54th Battalion had a policy at that time for reaction to a convoy ambush. The drivers of the "Triple - 6" (666th) had learned to get out of the trucks and return fire, but they had no other choice. The disabled trucks blocked the road ahead of them. Stopping turned out to be a bad idea. Drivers climbed out of their vehicles and returned fire while NVA swarmed down over at least three trucks placing satchel charges on the cabs.[22] J.D. jumped out and took cover between his truck and the hill side. A convoy halted in the kill zone was exactly what the enemy wanted. Since the drivers were support troops, they did not carry much ammunition and it quickly ran out.[23] Fortunately, someone drove along the line of damaged trucks and picked up wounded drivers.[24]

The highway patrol from 504th MP Battalion was notified and arrived on the scene while the ambush was still in progress. SGT Buford Cox, Jr., Woodell, Morin, Bledsoe and Phipps found the drivers stopped and milling around, blocking traffic.[25]

CPT Geise called the 1st Air Cavalry for support. In ten minutes the enemy had destroyed or damaged 30 vehicles, killing seven men and wounding 17. Helicopter gunships from A Troop, 1/7th Cavalry arrived and began searching for the enemy.

SGT Cox stopped an officer in a jeep and asked where he was heading. The officer said he was going back to Pleiku for help. Cox told him the best thing he could do was go to the front of the line of trucks and lead them out of the kill zone to An Khe. Cox "forcefully encouraged" the officer to do so and went with him. Cox found that the drivers had dismounted their trucks and were trying to return fire. He ordered every driver he could find to get back in their trucks and drive them through the kill zone. This probably saved a lot of lives. Once the drivers understood they needed to clear the kill zone, Cox and his men began to lay down suppressing fire with the machine guns, M-16s and the M-79 grenade launcher that Cox always carried in his jeep.

At approximately 1930 hours, the 3rd Platoon, C Company sent reinforcements in two gun jeeps with SGT Tomlin, Midolo, Desena, Leclair, Carter and Palumbo. They arrived at the scene and established security and traffic control, and directed recovery vehicle operations. At 1940 hours a second reinforcement of two more gun jeeps with 1LT Thomas Briggs, Melnick, Young and Trumbo performed a recon to the west of the ambush site to determine the position of the last damaged vehicle. When they arrived at kill zone it was dark and the gunships had left. Wrecked trucks were everywhere as well as the bodies of dead. A Transportation Corps officer wanted to know if there were any other vehicles that needed recovered so Briggs took two jeeps and drove down the road. Damaged trucks were arrayed at a variety of angles to a straight line down the middle of the road. Briggs remembered the body of one American soldier, cut in half at the waist, draped over the hood of a truck. Most of the damage was to the lead vehicles in the convoy.[26]

LTC William K. Hunzeker, Commander of the 34th Supply and Service (S&S) Battalion, took charge of the recovery operations.[27] 2LT Burrell Welton was leading a fuel convoy of the 359th Transportation Company that came up and halted a couple miles behind the 8th Group convoy just as it had been ambushed. Helicopter gunships were already shooting up the hill side and it was getting dark. He learned some the drivers had either left their weapons in camp or wrapped them up under seat to keep them from getting dirty. So the enemy had stood on running boards shooting inside the cabs. The 359th drove through the kill zone while the 34th S&S Battalion was cleaning up the mess. The trucks were stopped bumper to bumper. Either they had too tight an interval going into the kill zone or had bunched up when trucks stopped ahead of them, because many drivers would not drive in the dirt around the damaged truck for fear of driving over buried mines.[28]

The AC-47 gun ship, "Spooky," arrived at 2020 hours, but the enemy had long escaped under the cover of darkness.[29] All Military Police remained on the scene until 0130 hours, when all vehicles except one were either removed to Camp Radcliff or to a position adjacent to an artillery firebase near CP 89. Afterwards,

A convoy was late coming back from Pleiku and was attacked west of An Khe by VC pulling boards in front of the trucks with anti-tank land mines on them.

Damage done by satchel charges.
Photos courtesy of Walter Medley

Vietnam War 23

Detail of damaged trucks. Photos Courtesy of Walter Medley

1LT Briggs received approval from LTC Hunzeker to send all MPs except two armored jeeps back to An Khe, since the 70th Engineer Battalion was on the scene with wreckers and gun jeeps. The last highway patrol remained until 0300 hours when the last vehicle was hauled to the artillery firebase. The 1st Cav pursued the enemy for about a week estimating their strength was around 60.[30]

Three of the killed, SSG Claude L. Collins, SP4 Ronald W. Simmons and PFC Arthur W. Reinhart, were from the 512th Light Truck. PFC Roy L. Greenage and PVT Lloyd R. Hughey were from the 669th, PFC William A. Gunter was from the 523rd, PFC Robert L. Stebner was from the 57th Transportation Companies.[31] Six of the 17 wounded came from the 669th and only one returned to duty. The others were evacuated to the United States for further treatment.

This ambush sent shock waves throughout the 8th Group. For the truck drivers the nature of the war had changed significantly. They had become the primary objective of the enemy offensive and from then on when they drove out the gate the drivers expected that they could be killed.

NVA Lessons

The NVA tactics had worked. Hitting an empty convoy returning from Pleiku did not shut down the supply line but only reduced the number of vehicles and drivers available for line haul. This had apparently been a rehearsal, as the NVA had deliberately planned the ambush for late that evening so they could escape under the cover of darkness. From their success and the US reaction, they developed their plans for future ambushes. For two years the NVA had sparred with the air cavalry only to learn to avoid American tactical air power. The speedy response of this tactical air power made the difference between the outcome of this ambush and the annihilation of French Mobile Group 100. The NVA would limit the duration of their ambushes to about ten minutes which was short of the arrival of helicopters or the AC-47. The NVA would take two months to plan, rehearse and execute their next convoy ambush.

Lesson

There was nothing obvious to predict that the enemy would change his tactics and target the convoys. An analyst might have drawn that conclusion by looking at the big picture. The insurgents had not any successes on the battlefield against the air assault units. He therefore had to find another weakness. The dependency on fuel hauled by trucks was a weakness. It was clearly known that this enemy was proficient at ambush tactics and had targeted convoys.

Surprise gave the enemy the advantage in the initiation of the ambush and after that it became a contest of the employment of fire power but the drivers were outgunned. LTC Melvin M. Wolfe, 8th Group XO and former Commander of the 54th Battalion, came up with the idea in the summer of 1967 to experiment with gun trucks and LTC Philip N. Smiley, Commander of the 27th Battalion, built sandbag pill boxes on the back of two 2½-tons, which unfortunately were not in this convoy when it was hit. The only alternative to gun trucks was training the drivers to fight as infantry.

Since the enemy had begun targeting the convoys, they needed protection. The question was whether the protection was the responsibility of the combat unit which had responsibility for the area or the truck companies themselves. Complacency had set in. It was not a question of whether the SOP for reaction to an ambush was adequate as none of the drivers knew it. They had never had a need to. The 8th Transportation Group would have to look at both active and passive measures to protect the convoys.

Doctrine Change

By US Army doctrine, the ground combat commander has responsibility for the routes the convoys pass through, which means the ground commander, or battle space owner, has responsibility to keep the roads safe and open. But no matter how the ground commander disperses his forces, they cannot cover everything, and the enemy will attack in the weak area. Military Police likewise have responsibility for route security, to patrol the roads looking for signs of kill zones, but in spite of their effort, ambushes still happen.

These trucks show shrapnel damage from the left side of the road. Photos courtesy of Walter Medley

Vietnam War 25

Both of these forces can provide a reaction force once a convoy is ambushes, but the immediate survival of the convoy depends upon internal security measures.

External Action Taken

Within a week, LTG Stanley R. Larsen, Commander of I Field Force, Vietnam, held an informal meeting at the 1st Cavalry Division's Headquarters at An Khe. This was not to assess blame but to determine what measures should be taken to protect the convoys. Doctrinally, route security was the responsibility of the unit responsible for the area. As it turned out, everyone had become complacent. They reviewed each unit's responsibilities in reference to convoy security. Since the 1st Cavalry Division had most of its units in the field, it could not guard the road. Larsen instead ordered the 4th Infantry Division (Mechanized) at Pleiku to secure the road.[32]

They did so by setting up tanks and APCs at more check points along the road. These check points were usually located at bridges and culverts or likely trouble spots. There were bridges and culverts about every

COL Joe Bellino, 8th Group Cdr examining battle damage.
Photo courtesy of the US Army Transportation Museum

As a result of the convoy ambushes, the I Field Force Commander assigned tanks and armored personnel carriers at security check points along QL19.
Photo courtesy of the US Army Transportation Museum

three US miles. From these check points the security forces could serve as reaction forces in the event of any nearby ambush. Neither the Koreans nor the Americans, however, could station mechanized troops in the most likely ambush locations in the mountain passes.[33]

In the event of an ambush, someone with a radio would call, "Contact, Contact, Contact" and all combat units in the area would be at the convoy's disposal. The 4th Infantry Division (Mechanized) had tanks and M113 armored personnel carriers stationed at check points along QL19.

Since the security forces had responsibility for their section of the road, they had command and control of the convoys entering into their area of operations. They could stop the convoys if they detected trouble up ahead. They also passed on current enemy intelligence about the road. If a conflict arose between the convoy and the security force then the highway coordinator of the Traffic Management Agency would resolve it. This command and control relied on radio communication. Convoy briefings every morning hopefully included the accurate radio frequencies of the security forces guarding the road. Convoy commanders had to call in and authenticate to the frequency of the next security force at the top of An Khe Pass. They soon ran out of range of their battalion headquarters when they passed over the mountains. They had to rely on the combat units to relay any messages back to battalion headquarters. They next authenticated to the frequency of the force at An Khe. There the convoy halted to drop of supplies destined for those units. The next frequency change was with the unit at Mang Giang and at last the convoy commander switched to the frequency of the force based at his destination at Pleiku.[34]

Larsen felt that the trucks should not have been out on the road at night. He ordered the road closed at 1515 hours for eastbound traffic out of Pleiku and at 1700 hours in the evening out of An Khe instead of 1900 hours. To depart on time, the convoys left Qui Nhon earlier. They rolled out the gate at 0300 hours. The 815th Engineer Battalion also began clearing away the vegetation back 1,000 meters on both sides of the road with heavy grading equipment called "Rome Plows." This measure hoped to deny enemy the cover of the jungle to hide in.

The Military Police also provided route security. B Company, 504th Military Police (MP) Battalion cleared the road with two gun jeeps armed with M-60 machine guns each morning from An Khe to Check Point 102 below Mang Giang Pass, then the gun jeeps of C Company, 504th would escort the convoys the rest of the way into Pleiku. The MPs would send an escort off two gun jeeps out. One would lead ahead of the convoy and the other would follow behind at a distance. Other than that, no combat vehicles would escort the convoys. On occasion, convoys could have access to occasional air support. When intelligence reports indicated likely enemy activity, an L-19 "Birddog" observation plane would fly surveillance over the area. After that meeting, both LTC John Burke, Commander of the newly arrived 124th Transportation Battalion and acting 8th Group Commander, and Wolfe realized that their convoys still had to defend themselves.[35]

Internal Actions Taken

Internally, complacency had set in with the drivers of the 8th Group since no convoy had been ambushed during the first two years of the ground war. The first thing that LTC John Burke, acting 8th Group Commander, did after the ambush was read the standard operating procedures (SOP) and make any needed changes. As engineers paved the road toward Pleiku, the trucks could drive faster than 35 miles per hour. The official 8th Group SOP required convoys to obey speed limits and reduce speeds commensurate with road, weather and traffic conditions. It stated, "Speed generates carelessness." Speed limits through villages reduced to around 15 miles per hour. The truck drivers knew it was harder to hit a fast moving target. Some convoy commanders briefed that trucks should drive as fast as they could go.[36]

It then became imperative that trucks maintain a 100-meter interval between trucks. This interval limited the number of trucks in the kill zone. Under these conditions, the 1,000 meter kill zone of the ambush on 2 September would have only caught ten trucks instead of 37. The ambushes occurred at places where traffic had to slow down. Drivers should also watch for changes in familiar scenes along the route. There were usually changes in behavior to indicate an ambush up ahead. The absence of people on the streets, the gathering of unusual looking people or even civilian vehicles parked alongside the road waiting for the convoy to pass indicated that there was danger ahead. The locals knew very well what the enemy was up to in their area.[37]

In the event of an ambush, the thin skinned vehicles had to rapidly clear the kill zone. Truck drivers would not stop in an ambush for any reason even if wounded. Those that could drove out of the kill zone, those that could not turned around and drove back to the security of the nearest check point. If the vehicle was disabled then the driver should pull off to the side of the road, dismount and jump on a passing vehicle. If the disabled vehicle could not pull off to the side of the road, then the next vehicle would push it out of the way. If the task vehicles could not turn around then they would halt at 100-meter intervals, the drivers would dismount and provide security. From then on the convoy commander briefed these procedures every morning. However, these were just passive measures to limit vulnerability.[38]

Gun Trucks

LTC Burke then met with his battalion and company commanders to discuss what active measures the 8th Group could do to protect their convoys. Since they could not count on the combat units to protect them, they needed their own offensive firepower. The 8th Group initially borrowed M-55s, Quad .50s mounted on M-35 2½-ton trucks, from the local artillery unit. The Quad .50 was four synchronized .50 caliber machine guns mounted on the bed of a 2½-ton truck. The Quad .50 gun trucks, however, required a crew of six, one driver, one gunner and four men to reload each of the guns. The crews took their training from the artillery unit that provided the weapons.[39]

Since each company had a few 2½-ton trucks for

First gun truck design with wood and sand bags. 6th Trans BN in Long Binh. Photo courtesy of the US Army Transportation Museum

Inside of gunbox created with sandbags and wood walls. M60s were hand held. Photo courtesy of Ted Biggins

administrative duties, they converted them to gun trucks within weeks of the 2 September ambush. In World War II and Korean War, a gun truck was any truck with a ring mounted machine gun. As a task vehicle they could only return fire while fleeing the kill zone. The battalion commanders directed the construction of six gun trucks per company and the company commanders picked the crews. The crew generally consisted of a driver, NCO in charge (NCOIC) armed with an M-79 grenade launcher riding "shotgun" in the passenger seat of the cab and two gunners with handheld M-60 machine guns standing in the gun box. When ring mounts arrived, some crews added an M-2 .50 caliber to the cab. The new concept of a gun truck was a dedicated weapons platform which could maneuver to protect task vehicles.

The gun trucks were integral to the companies and the leaders appointed the crew members. Only a few crews had named their gun trucks at that time. That early in their development, they were very generic and had not taken on any character. The new SOP called for a gun truck ratio of one for every ten task vehicles. Their mission was to move to the flank of the kill zone and return fire on the enemy. Up until 24 November, the gun truck doctrine had not been tested in an enemy ambush.

The limitation of the gun truck was the availability of radios. Few trucks had them and then only one. Usually just the convoy commander and the assistant convoy commanders had radios so the gun trucks without radios had to watch them for instructions. This required a gun jeep with a radio to drive nearby, either ahead or behind the gun truck. The key to a coordinated attack is communication and without radios, gun trucks' actions were independent of each other. With the proper 100-meter intervals, long convoys could stretch over a mile and some gun trucks may not even hear the gun fire to respond.

The new 2½-ton gun trucks were initially hardened with sandbags and wood, which quickly fell apart from the beating they took on the rough road. Fortunately, precut steel plating that had been ordered during the summer arrived shortly after the 2 September ambush. The steel plates were rectangular in shape with a series of precut holes for bolting the plates together. The plates had small portals cut in them for gunners to shoot through, but most preferred to fire over the top of the box with an unobstructed field of vision. There were steel plates for the doors and shield with view portal that could be flipped up to protect the windshield in the event of ambush. The armor was painted OD to match the color of the trucks, and out of pride some crews named their trucks and stenciled the names on the sides.

Clockwise: **Bounty Hunter and Nancy were Quad .50 gun trucks. 2¼-Ton M35A2 truck cargo, with quad 50 machine gun. Brand new factory kits were added to trucks in Vietnam.**
Photos courtesy of the US Army Transportation Museum

Validation
11 and 24 November 1967
54th Transportation Battalion

On 11 November 1967, 1LT John M. Arbuthnot II, of the 585th Transportation Company, led the lead serial of an eastbound convoy from Pleiku back to Qui Nhon. A gun truck led each serial of ten task vehicles. By this time each gun truck had a minimum of two handheld M-60 machine guns in the gun box and possibly a ring-mounted .50 caliber machine gun over the cab. Most 2½-ton gun trucks had a single wall of plate armor around the bed of the truck as a gun box. Some gun trucks were Quad .50s. The drivers were armed mostly with M-14s and a few units with M-16 rifles with four or five 20-round magazines. SPC Brown, a driver on a gun truck of the 666th, initially carried only twelve 40mm rounds until he learned how fast he could expend them in a fire fight. So this early, the crews were not carrying a lot of ammunition.[40]

About two miles west of the 1st Cavalry Division Check Point 89 in the same area as the 2 September ambush, Arbuthnot's convoy came under fire at approximately 1515 hours. The enemy initiated the ambush with two to three command detonated mines planted on the south (driver's) side of the road, which did no damage, and was followed by 100 rounds of small arms fire from between 10 to 15 guerrillas also concealed on the south side of the road. As soon as the call, "Contact, Contact, Contact," went out the 1st Cavalry Division launched three helicopter gunships and a reaction force. Since the mines failed to stop any vehicles, the fire fight lasted for about 30 seconds, concentrated on the last vehicles of the serial. The drivers did as taught, drove through the kill zone and assembled at Check Point 89, where they realized the damaged vehicle had five rounds in the radiator, one in the right front tire, and one hit the armor plating on the driver's door. The damaged vehicle was taken to An Khe for repair overnight while the rest of the convoy

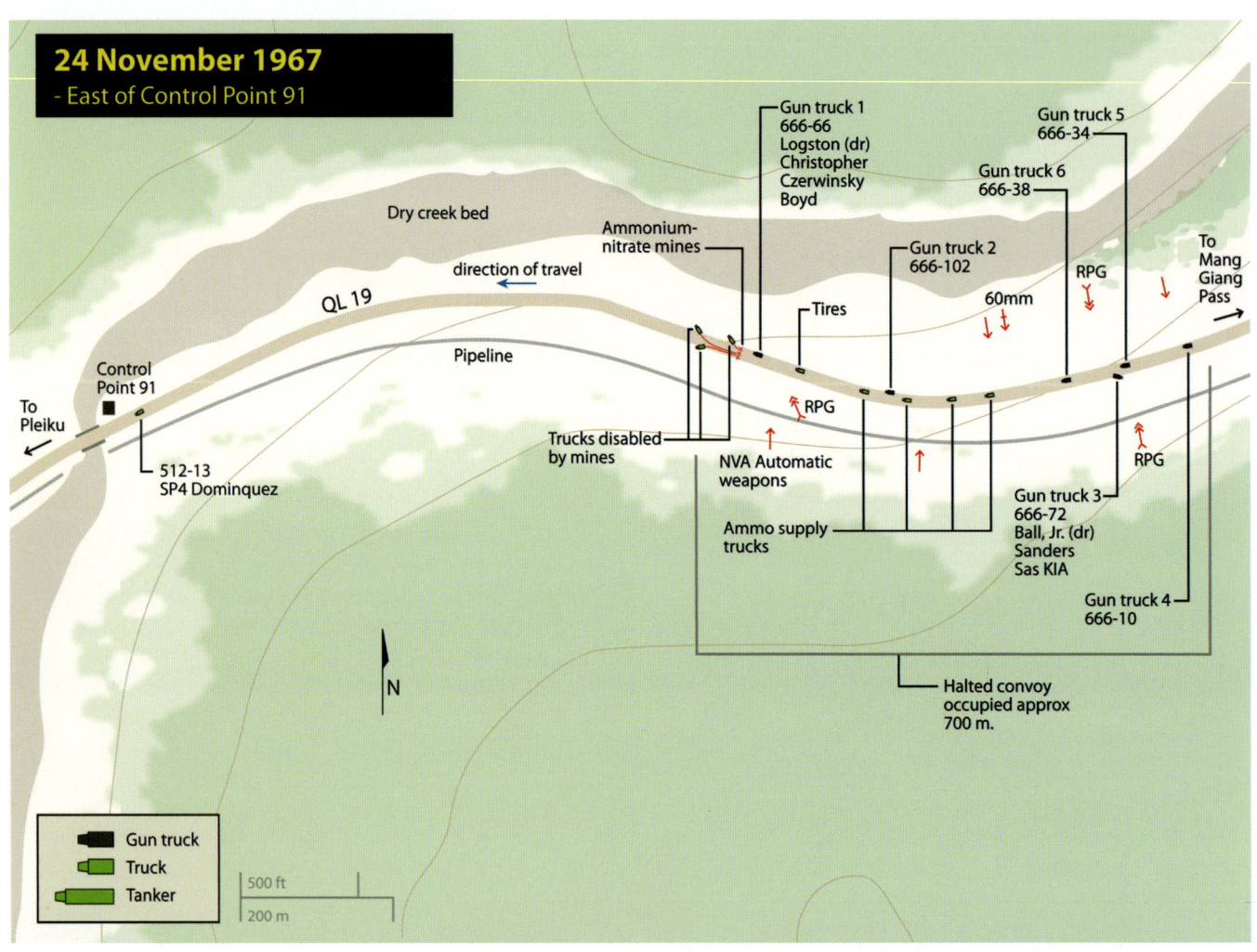

View of kill zone as convoy entered it with disabled vehicles on right side of the road several days later.
Photo courtesy of Bill Eichenberg

continued on to Qui Nhon. The helicopters arrived within five minutes and the 1st Cav troops arrived in ten to conduct a sweep of the area. The enemy was just warming up.[41]

On 24 November 1967, approximately 70 NVA launched its next large scale ambush on a 54th Battalion westbound convoy. 1LT James P. Purvis, of the 666th, led a westbound convoy of forty-three 5-ton cargo trucks, fifteen 2½-ton trucks and a maintenance truck under the protection of six gun trucks and three gun jeeps. It was divided into six serials of about ten task vehicles and one gun truck per serial. The M35 2½-ton gun truck and M35 2½-ton cargo trucks of the lead serial belonged to the 666th Transportation Company.[42]

Jerry Christopher, of the 666th Transportation Company, rode shotgun in the cab of the lead 2½-ton gun truck, which traveled 20 miles per hour up QL19 toward bridge over the dry creek bed about 19 kilometers east of Pleiku. Tall grass covered the ground to the edge of the wood line 50 to 100 meters from the road. About 1000 hours, the third gun truck of the 666th, 17th in line of march, the driver, SP4 Gipsey B. Ball, Jr., pulled out of line because a tire was going flat. Acting sergeant, SP4 Arthur J. Hensinger drove up in his gun jeep and told them to get back in line since the tire was not that bad and they were almost at Camp Wilson. SP4 Roy A. Sanders and PFC Robert L. Sas were standing up behind the cab looking around with their handheld M-60s when Sas tapped Sanders on the shoulder and suggested they sit down since the convoy neared its destination. As soon as they took their seats, they heard an explosion.[43]

At 1005 hours, Christopher spotted ten paper bags spaced across the road and recognized them as fertilizer mines. He shouted to his driver, Bob Logston, "We're in the kill zone! "What?" Logston shouted over the roar of the engine. "We're in an ambush!" Logston floored the gas pedal and grabbed his rifle. The two machine gunners in the box opened fire with their M-60s. A B-40 rocket then slammed into the front end blowing off the left tire and part of the wheel. The gun truck slid to a halt 25 yards short of the mines. Christopher yelled into the radio, "Contact! Contact! Contact!" He tumbled out of the vehicle with Logston behind him and started firing his M-79 grenade launcher. Enemy fire ripped through the windshield, the engine block and into the armor plating on the side of the cab.[44]

The SOP for an ambush was for those vehicles in the 300-meter long kill zone to not stop but drive out. The next 5-ton loaded with small arms ammunition, down shifted, pulled out of line and roared around the damaged gun truck unaware of the daisy chain ahead. The mines blew off the front end and the truck swerved out of control off to the right side of the road. The third driver also accelerated his rig and ran over the remaining mines losing both front wheels. His truck slid 75 yards down the road and ended up in a ditch across the road with his load of 155mm high explosive projectiles on fire. SP4 Dick Dominquez, of the 512th Transportation Company, revved his engine and raced for the gap in

Vietnam War 31

Disabled trucks on the right side of the road. Photo courtesy of Bill Eichenberg

the road. Carrying a load of CS gas, he safely squeezed through the gap and headed on to Pleiku. His was the only truck to escape the kill zone. The enemy hit the next truck loaded 155mm projectiles with a B-40 rocket and set it on fire. It slid to a halt 50 yards from Christopher's gun truck. QL19 was completely blocked by damaged vehicles.[45]

Christopher began firing his grenade launcher at the suspected position of the B-40. The artillery ammunition load began to cook off. Each blast rocked the corpse of the gun truck near it. Christopher crawled to the front of the vehicle looking for his driver. Logston had been hit by machine gun fire below the waist and was a bloody mess. He was trying to crawl out of the firing line. Christopher called out, "Bob! Y'all right, Bob?" Christopher then pulled his driver into the elephant grass. Jerry asked, "What're we gonna do now, Jerry?" He looked up and saw helicopters circling high above. "Why don't they do something? Why don't they help us?" These were command and control birds with senior officers. Christopher pulled Jerry under the gun truck and bandaged his wounds. "Hang on – we'll make it OK." That was as much his wish as reassurance.[46]

Another rocket hit the tail gate above Christopher sending a shower of fragments all over SP4 Czerwinsky, a machine gunner. The other machine gunner, Jim Boyd, was hit in the arm. Both M-60 machine guns were smashed. While Christopher tried to save Czerwinsky's life, Boyd searched for a rifle and started firing away with his good arm. Christopher then saw an NVA sapper in the grass across the road. He fired with his M-79 not sure if there was enough distance for the round to arm. The round exploded on target.[47]

Sas and Sanders had heard explosions followed by gun fire up ahead and then around them. They saw enemy soldiers popping up and shooting at them from the left side of the road so Sanders and Sas then returned fire. Suddenly they heard a loud explosion on the passenger side door and then their gun truck started slowing down as cargo trucks began to pass them. Sanders shouted at Ball, but he did not answer, so Roy pulled back the canvas flap on the back of the cab and did not see his driver behind the steering wheel. He looked down and saw Ball lying on the floor on the passenger side. Ball yelled that he was wounded. When the ambush began, a B-40 rocket had passed just inches behind the second gun truck. The next rocket hit the cab, wounding Ball and throwing him to the floor boards. Sanders helped Ball back up behind the wheel. Both were scared and Ball wanted to turn the gun truck around, but Sanders told him they could make it through the kill zone. So Ball put pedal to the metal and Sanders returned to his duties on the M-60.[48]

Ball was racing through the kill zone and about to change gears when an RPG hit his driver's side door. The truck lumbered out of control to the left side of the road knocking Sas and Sanders to the bed of the truck. When the truck hit the ditch it flipped over throwing the crew around inside the box. The truck stopped upside down in the grass right next to the enemy. When Sanders regained his senses, he looked up at the spinning rear wheels of the truck. The gun truck armor was pressing

Gun truck driven by SP4 Gipsey Ball after it had been rolled right side up. Photo by courtesy of Geroge Marquez and James Lyles

down on his left thigh and he had no weapon, so he began clawing the ground to dig his leg out while the enemy shot out the tires above him. Sanders dug himself free with his hands only to find that his leg was broken and he had no weapon. He could hear his driver trapped in the cab crying for help and told him to be quiet since he heard enemy on the other side of the truck. Sanders saw Sas pinned under the door armor and crawled over to him only to discover he was dead. Looking for a weapon, he crawled to the front of the truck and saw an M-60 inches away, but an NVA machine gunner on the other side kept him from reaching the machine gun. He then picked up two fist-sized rocks and threw them hoping the enemy would think they were hand grenades giving him just enough time to reach the M-60. No sooner had he started crawling to the M-60 he heard a "plop" beside him. He then rolled away from the grenade when it exploded and threw him ten feet away. With fragments in his legs, he staggered to his feet and tried to get back to the truck when he blacked out.[49]

5,000-gallon fuel tankers in the first serial had burst into flames spilling their flaming contents down the road for 700 yards. Pallets of ammunition on the backs of the other trucks began to cook off. NVA sappers ran up to the vehicles, climbed atop and placed demolition charges on the cargo then fired down on the drivers hiding in the grass along the side of the road. Drivers returned fire knocking the enemy off of their trucks into the wreckage littering the road.[50]

Enemy fire also hit the gun truck in the third serial and damaged it. Ted Ballard, of the 523rd Transportation Company, was assigned to ride shotgun with a replacement and rode with his M-14 between his knees. Up ahead he heard what he thought was the sound of artillery then saw big clouds of smoke from behind a little rise of ground. He had never seen black smoke in that area before which raised his suspicions. Then the trucks ahead of his began to stop and the drivers jumped out on the right side of the road. Ballard realized he was in an ambush and jumped out on the right side. His driver jumped out on the left side and Ballard never saw him again.[51]

Ballard then heard small arms fire. The side of the road had been cleared of all vegetation to include grass by engineer bulldozers but the dirt and brush was dozed into small piles that offered cover for the enemy. Ballard saw what he thought were ARVN soldiers just walking toward the convoy, and then the other drivers started firing at them. When the Vietnamese returned fire, Ballard started firing at them and emptied two magazines. He carried 200 rounds, all tracers. The Vietnamese then ran into the convoy. The gun truck was up ahead and killed most of the enemy. When the ambush was over, Ballard's driver never returned, so he climbed in another truck. When they drove down the road, he saw about a half a dozen dead enemy by the side of the road where the gun truck had been. Further down, he saw that the black smoke was from burning tires.[52]

Further back down the road, a grenade had damaged the gun truck in the fourth serial leaving only the last two gun trucks undamaged. Meanwhile a tank and two APCs from 1st Platoon, C Troop, 2/1st Cavalry

Vietnam War 33

of the 4th Infantry Division guarding a bridge at Check Point 91, about a mile away, raced into the kill zone as soon as they heard shooting. Upon their arrival, the cavalry men saw burning trucks along the road with NVA soldiers climbing on them and hurriedly unloading the cargo. Upon seeing the approaching tracked vehicles, the enemy started to run but the cavalry men opened fire. The subsequent heavy fighting raged for another 20 minutes before the helicopters and jets arrived, while pallets of ammunition on the trucks exploded. The troopers loaded Sanders and Ball into their APC and then drove them to an awaiting medevac helicopter. One helicopter landed and General Creighton Abrams stepped out. The combined gun trucks, cavalry and air support concluded the fight after another hour and Thomas Weston's APC had fired off nearly all 30,000 rounds of ammunition. A sweep of the battlefield counted 41 enemy dead and captured four wounded. The convoy had 14 trucks and four gun trucks damaged and six to nine trucks completely destroyed. 17 drivers were wounded and three were killed, SP4 Arthur J. Hensinger, SP4 Charles E. Metcalf, and PFC Robert Sas. General Abrams, an old WWII tanker who had been watching the battle from above, praised the truck drivers for fighting like tankers.[53]

Lesson

Clearly the fire power of the gun trucks caught the enemy by surprise. Assaulting the convoy as they had before made them pay a high price, yet the convoy also sustained high losses. The NVA would not be so reckless the next time. They would conduct convoy ambushes nearly every week and test the capabilities of the gun trucks. With a limited arsenal of weapons available, the enemy would vary his tactics trying to determine what weapons to initiate with, how large should the kill zone be and what vehicles to destroy first. In spite of tanks and mechanized infantry at check points, the enemy would conduct ambushes at the areas not covered, most likely in the mountain passes. A short stretch of road below Mang Giang Pass would become known over the next few months as "Ambush Alley." An Khe Pass would be the next favorite ambush area. During the last two convoy ambushes, he had initiated the ambush on the lead vehicles. The 8th Group, however, was on the right path to a solution. An ambush is a quick and violent attack that relies on surprise to be effective. The gun trucks would learn that they needed to turn the fight around fast with greater violence. The gun trucks needed more fire power. While the Quad .50s had far superior fire power, each of the four guns required a loader. This required a crew of six; a driver, a gunner and four loaders, far more than the truck companies could afford. The Quad .50s would soon fall from the inventory. The box style gun trucks began to mount M-2 .50 caliber machine guns either on ring mounts over the cab or on pedestals in the gun box. The 7.62mm rounds of the M-60 did not have the penetrating ability of the .50 caliber. Eventually, three machine guns became standard for each gun truck.

Not having radios in each truck prevented the lead gun truck from warning the others of the mines across the road.

Pre-Tet Offensive

The NVA commanding general, Vo Nguyen Giap, was confident that the conditions were ripe for the Viet Cong to rise up and join the invading NVA in the final phase of his insurgency strategy. The official commencement of offensive operations began on the night of 31 January 1968, the Vietnamese Lunar New Year celebration known as Tet. The NVA had agreed to a cease fire during the holiday so that the Army of the Republic of Vietnam (ARVN) soldiers could go home for the holidays. In effect, the NVA hoped to catch the ARVN completely off guard. The only force that could seriously contend his operations was the Americans. Giap clearly understood the American dependence on supplies and knew that he had to sever the American supply line hoping to starve the combat units spread throughout the Central Highlands and Highland Plateau. For the 8th Transportation Group, the Tet Offensive began with a convoy ambush on 24 November 1967. From then on large scale convoy ambushes would become weekly occurrences with mining and sniping taking place daily. The objective of the enemy convoy attacks was to completely shut down the supply line.

It took the enemy usually a week to plan and rehearse large scale ambushes, so with the greater frequency of ambushes, the gun trucks and enemy sparred with each other to find what worked. The

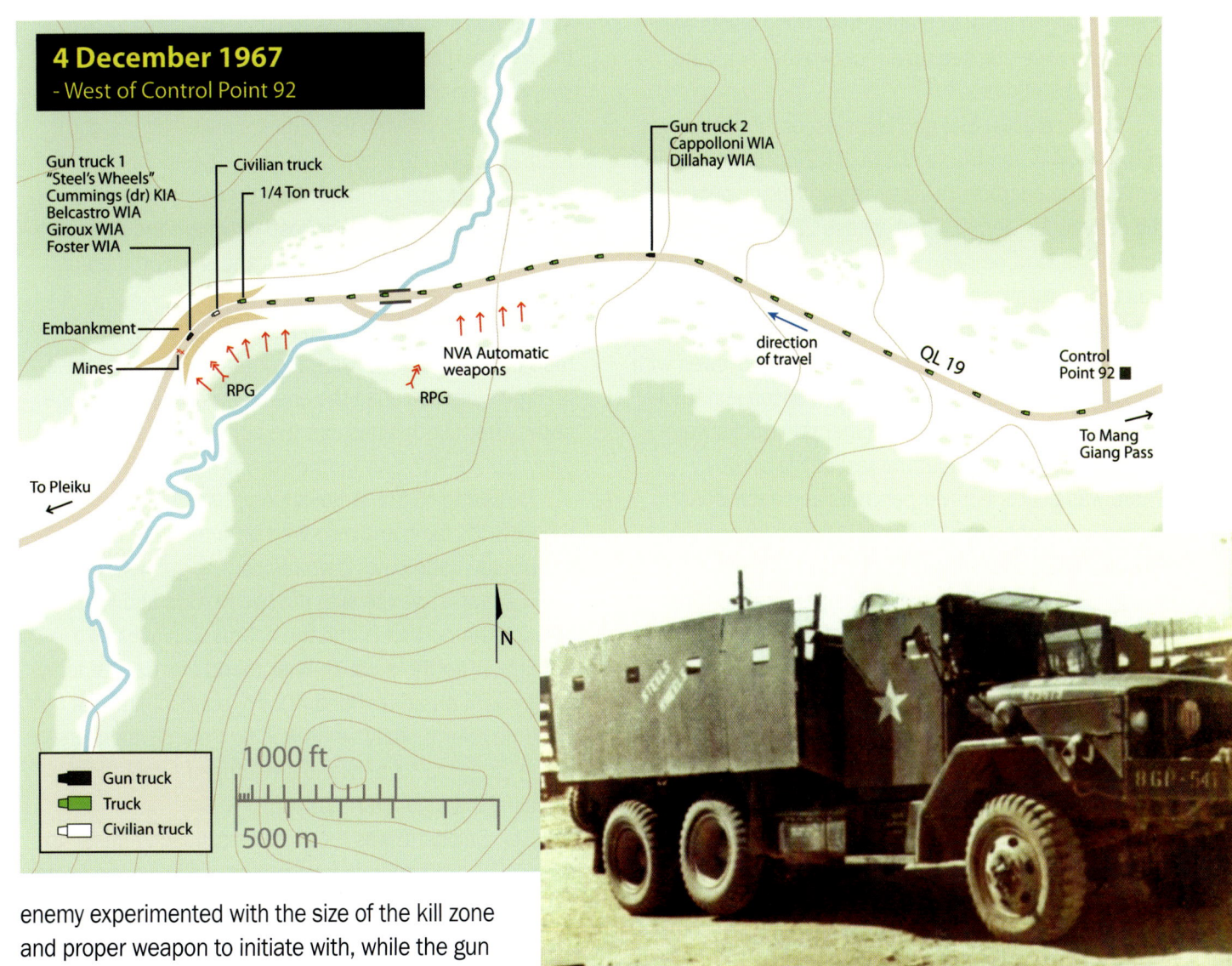

Steel's Wheel as it looked when the name was painted on it after the ambush. Photo courtesy of James Lyles

enemy experimented with the size of the kill zone and proper weapon to initiate with, while the gun trucks and crews developed their tactics and proper way to react to contact.

4 December 1967
54th Transportation Battalion

The 669th Transportation Company had built three gun trucks after the September ambush. 1st and 3rd Platoon bolted or welded the precut ¼-inch steel plates to the beds and doors of 2½-ton administrative vehicles, while the 2nd Platoon mounted their armor on what was probably the first 5-ton gun truck. SGT Dennis J. Belcastro was the NCOIC of the 1st Platoon gun truck and called it, "Bel and the Cheese Eaters," and 3rd Platoon called theirs "Steel's Wheels" after their platoon leader, 1LT James R. Steel. The 2nd Platoon gun truck had no name yet.[54]

SGT Belcastro was supposed to have 4 December off, but early in the morning someone woke him up to fill in for the NCOIC of the 3rd Platoon gun truck. SP4 Harold Cummings was the driver. According to the new policy, Belcastro would ride in the gun box with the two M-60 gunners, Frank W. Giroux II and Joseph Foster. That was the first week NCOICs rode in the gun box, so they could augment and direct the gunners' fire power. Belcastro was armed with an M-79 grenade launcher, .45 Colt semi-automatic pistol and M-14. They carried 30 100-round cans of M-60 ammo, two dozen 40mm rounds, and Belcastro had four 20-round magazines taped together in tandem so he could quickly change magazines. They stored the extra ammunition in a metal box, three feet tall, four feet wide and three feet deep. To access it, one had to lift the lid. Sandbags lined the

Vietnam War 35

floor of the gun truck to protect from mine blasts and sandbags were stacked up about two feet along the inside of the gun box.[55]

Six gun trucks, four gun jeeps escorted this convoy of 58 M-54 5-ton trucks, 11 M-35 2½-ton trucks and a maintenance truck. Belcastro's gun truck would lead the convoy. 1LT Jerry L. Todd, whose 2nd Platoon, 57th Transportation Company was attached to the 669th was the assistant convoy commander and would follow behind Belcastro's gun truck. One commercial truck and 10 M-54 5-ton cargo trucks of the 669th followed with SP5 Dennis Cappolloni's 2nd Platoon 5-ton gun truck leading the second serial. The other gun trucks were evenly spaced throughout the rest of the convoy. There was only one driver per truck and each was armed with either an M-14 or M-16 with four to five 20-round magazines.[56]

At 0815 hours, the convoy snaked through the winding road of the 2 September kill zone, passed the security force at Check Point 92 and approached the base of Mang Giang Pass. The ground was hilly covered with two-foot high grass with dirt mounds 100 yards on each side of the road. A dense jungle flanked the grassy field. As Belcastro's gun truck rounded a bend, the road was cut through hill with a 15 to 20-foot embankment on right side and 25-30-foot embankment on the left side. Small arms fire came the driver's side (south) of the road and about 30 seconds into the fire, Cummings, slammed on the brakes and yelled back the enemy was dragging a board with three mines pulled across the road. Belcastro told him to keep going and drive around it. No sooner had Dennis finished the sentence when an RPG fired from the edge of road hit where the door met the windshield killing Cummings and stopped the gun truck in the road.[57]

The ambush opened up from the left side of the road but after the ambush, the infantry found satchels on right side indicating the enemy was set up to hit the convoy from both sides, but the convoy was 15 minutes early. So the NVA only hit the convoy from the port side.[58]

Belcastro called "Ambush, Ambush, Ambush," on the PRC-25 radio and also fired off a red flare to alert the trucks that did not have radios. The gunners laid down suppressive fire with the two M-60s and M-79 through the portholes and the enemy was right on top of them. After the ambush, they found one dead enemy soldier with an unexploded 40mm round in him that had not time to arm because he was so close. The drivers in the cargo trucks behind them jumped out of their vehicles and also returned fire.[59]

SGT Belcastro was hit when he exposed himself to open the box to get more ammunition. Giroux and Foster

SGT Belcastro's gun truck. Photo courtesy of Dennis Belcastro

36 Convoy Ambush Case Studies - Volume I

were also wounded during the fight. In spite of their wounds, the crew of the gun truck and Todd's gun jeep crew returned immediate suppressive fire. The enemy was a company-sized force and five minutes into the ambush, the enemy conducted two simultaneous attacks against the lead and second gun truck. Steel's Wheels was 200-300 meters behind the lead gun truck. Most of the damage was in the second part of ambush. The suppressive fire from Dennis' gun truck kept the enemy away from the trucks in his serial. Four cargo trucks received flat tires but the drivers returned fire breaking off the assault. Three minutes later the enemy made another assault, which the drivers also beat back.[60] Meanwhile the other four gun trucks raced into the 3,000-meter long kill zone kill zone multiplying the suppressive fire on the enemy. Gradually the enemy fire tapered off and the gunners kept the enemy at bay. Belcastro had run out of 40mm rounds and had to fire his M-14 and .45. Girouz and Foster had burned up three M-60 barrels firing off nearly all their ammunition.[61]

The helicopter gunships arrived at 0827 hours, 12 minutes after the call, "Ambush, Ambush, Ambush" went out. As soon as the reaction force heard the call of contact, it left Check Point 92 and arrived at 0830 hours. By that time the intense fire from the gun trucks had broken up the enemy ambush, killing 13 enemy soldiers and capturing one wounded at a loss of only one American killed and eight wounded. Besides the crew of the lead gun truck, SP4 Dennis Cappolloni and SP5 Robert Dillahay in Steel's Wheels were wounded along with the following soldiers attached to the 669th; PFC Gerald T. Hyatt, Ray E. Gordon and Terrance N. Maddox. All returned to duty. The loss of vehicles was one gun truck destroyed and one jeep and four trucks slightly damaged. The EOD teams found seven satchel charges on the bodies indicated they had planned to stop the convoy and place the charges on the vehicles as they had during the 2 September 1967 ambush. LTC Robert L. Runkle, Commander of the 1st Battalion, 5th Cavalry commended the officers and men of the 669th for the outstanding part they played in the ambush. SGT Belcastro received the Bronze Star Medal with V device for his actions. 1LT Tillotson, 1LT Steel, and the crew of Belcastro's gun truck earned the Army Commendation Medal (ARCOM) with V devices for their role.[62] 1LT Todd received the ARCOM with V devices for his coolness under fire and the way he directed the helicopter gunship. His M-60 gunner, SP5 Stanley A. Runnas, also received the ARCOM with V device for his deadly and accurate suppressive fire.[63]

Lesson
The enemy prepared what looked like a linear kill zone and wanted to stop then overrun the convoy as it had on 2 September. For some reason the satchel charges were found on the opposite side of the road from the assault force. Belcastro believed the enemy had no time to complete their kill zone. Again the enemy picked a bend in the road a safe distance from the security check point. The enemy had initiated the last two ambushes with a daisy chain of mines pulled across the road instead of destroying the lead vehicle with rockets. The truck companies had not anticipated this. Clearly the mines had prevented the drivers from clearing the kill zone as their SOP called for. Later, the gunners of the gun trucks learned to fire their machine guns at anything suspicious along the road. They preferred this to firing the grenade launcher or rocket from the Light Anti-tank Weapon (LAW) as the latter would definitely set off an explosion but not always detonate the mines. The tracer rounds of the machine guns could set off the mines and the secondary explosions would let the drivers know the path was clear.

Once the lead gun truck was stopped in the kill zone, the gunners and drivers of the cargo trucks managed to lay down enough fire to beat back two enemy assaults until the gun trucks from behind came up to increase the fire power. Suppressive fire power reduced the number of vehicles destroyed or damaged and increased the number of enemy casualties until the armored and air cavalry arrived to take over the fight.

From the point of view of the convoy, the 12-minute reaction time of the air support and 15-minute reaction time for the security force was too long. The gun trucks definitely needed to carry more ammunition. After that ambush, the crew of the 3rd Platoon gun truck arranged the ammunition box where it opened down into bed so no one would get exposed when he opened it. Because the enemy got so close the grenade rounds did not arm, Belcastro felt he should have had some 40mm buckshot rounds.

Vietnam War 37

15 January 1968
359th Transportation Company
(Petroleum, Oil and Lubericants)

The 359th Transportation Company (POL) was based out of Camp Ratcliffe at the top of the An Khe Pass. The fuel pipeline ended at An Khe where the fuel tankers of the 359th pushed the fuel to military camps at Pleiku and beyond. The M54 5-ton tractors hauled M131 5,000-gallon fuel tank trailers. As a POL Transportation Company, the 359th fell under the command of the 240th Quartermaster Battalion, but with the recent increase in ambushes of 8th Transportation Group convoys along QL19, the 359th likewise built one 2½-ton gun truck. They welded and bolted ¼-inch steel plate to the side and back of the bed, and on the doors of the cab. The armor plates had firing ports cut in them, but one plate was factory cut shorter than the other offering only protection up to the waist. They lined the floor of the bed and cab with sandbags and welded two pedestals in the back for M-60 machine guns. The pedestals allowed the gunner to swing the machine gun a full 360 degrees and freed one hand to feed the belt of ammunition. Although the M-60 was designed as a light machine gun, firing the machine gun in a standing handheld position would physically wear out the gunner. The company also armored a one ¾-ton truck with a pedestal mounted M-60 machine gun. The company experimented with a Quad .50 on the back of a 2½-ton truck on loan from Qui Nhon. Four .50 caliber machine guns firing in one direction would prove intimidating, but the requirement for four loaders in addition to the gunner took five drivers off driving trucks, and the fact the loaders had no protection from enemy fire

15 January 1968
- Ambush Alley west of Phu Yen

38 Convoy Ambush Case Studies - Volume I

motivated the 359th to return their Quad .50. Similar to 8th Group, each truck had only one driver each armed with an M-14.[64]

The company commander had asked SP4 Ron Kendal to drive the gun truck and SP4 Timothy Wheat manned one of the two M-60s. Tim had enlisted at the age of 17 and once he turned 18, he volunteered for Vietnam. He joined the 359th POL in March 1967 as a replacement for the original members who deployed with the company. Ron arrived the next month. The other M-60 gunner position was rotated by different truck drivers each mission. Each person carried his M-14 and spare magazines in a footlocker. The footlocker also contained two 4,000-round cans of 7.62mm ammunition linked in a continuous belt for a mini-gun, which worked fine for an M-60. Up until then the 359th had only received harassing fire so that was considered more than enough ammunition.

On 15 January 1968, the 2½-ton gun truck led a fuel convoy of 40 to 50 fuel tankers to Pleiku. 2LT Burrell Welton, 3rd Platoon Leader, was the convoy commander. The skinny OCS graduate did not look impressive, but all the Soldiers liked him and wanted Welton as their convoy commander. 19-year old Welton had dropped out of college and enlisted to be a helicopter crew chief until he realized he could make more money as an officer so he applied for Armor OCS and Ranger School and was accepted to both. He unfortunately lost hearing in one of his ears when he was shot in head while home on leave at Los Cruces, New Mexico, so he could not go combat arms. He then asked to transfer to Transportation Corps OCS at Fort Eustis, Virginia. He had volunteered for Vietnam to get out of Fort Hood, Texas and arrived in June 1967. The men liked Welton because he treated them like people and not enlisted. Welton got along with the men because he came from the same background and was the same age. He felt as long as they did what he told them he did not have to get the company commander involved. He liked going out on the road and the men trusted his judgment.[65]

SP4 George Anderson, fresh out of the motor pool, drove the lead truck for the convoy that morning. He had 75 days left in country and never let the new guys forget it. SP4 Robert Dye would leave soon after him with 89 days left in country. They lined up first but George confessed to Bob he was having mechanical problems

Typical design of 2½-ton gun trucks in the 359th Transportation Company with one plate shorter than the others. M-60s were mounted on pedestals.

Above: **2nd Platoon's gun truck would later be named "War Wagon."** ***Bottom:*** **War wagon stopped on the side of the road with VN kids selling Cokes.**
Photos courtesy of the 359th TC Vietnam Association

Vietnam War 39

with his truck. The 359th POL liked to get on the road for Pleiku early in the morning before the 8th Group convoys reached the top of the pass.[66]

Paul Marcum and then Vernon Bush were supply specialists on temporary duty from the 304th Service and Supply (S&S) Company. The 304th had arrived on 5 October and was tasked to send ten soldiers to the 359th to fill in until replacements could backfill the original drivers of the 359th that had arrived the previous year. Marcum and Bush had been friends since they joined the 304th when it formed at Fort Devens, Massachusetts about four months before their deployment. When Bush heard Marcum was one of the ten going to the 359th, he volunteered to trade places with another soldier to stay with his buddy. That day they wanted to get in the front of the line so they could drop their load first and get something to eat at Pleiku. When they inspected their trucks in the motor pool that morning, Paul's truck had a flat, so Vernon saved a place for him up front, right behind Dye's truck. Unknown to them, this would give them a front row seat to the ambush. Paul had a pet German Sheppard he let ride with him in the cab.[67]

The 359th staged outside the back gate at An Khe for about 10 to 15 minutes then took off for Pleiku as soon as the sun came up to get ahead of the convoys of the 8th Transportation Group coming up from Qui Nhon. 8th Group convoys ran slow so 2LT Welton would tell the lead gun truck, "As long as you can see a tanker in your rear view mirror keep hauling ass." The MPs set up check points at An Khe and Pleiku to time the passing convoys and calculate their speed. If it appeared they had violated the 35 mph speed limit, then they issued the convoy commander a speeding ticket. Welton usually placed the slowest vehicle up front to regulate the speed, but he received lots of speeding tickets. Fortunately, this did not worry his commander.[68]

Before reaching Mang Giang Pass, Anderson's truck was having difficulty and slowing down the convoy so Dye's truck passed him. For the first time in 250 convoys, Bob Dye's truck was the lead truck in the convoy. The convoy climbed up the steep mountains of the pass about 15 to 20 mph. Around 0900 hours, the lead element of the convoy had just cleared the top of the Pass. Because the trucks usually spread out climbing up the Pass they usually slowed down at the top to regroup the convoy. They did not expect trouble because the 173rd Airborne Brigade manned a security check point at the bridge a few miles away with a tank and infantry. The convoy rounded a curve to the left and then passed through the village of Phu Yen. The Americans had bad relations with village of Phu Yen because a driver ran over and killed a young child in that village a few weeks before. The driver had not seen the child because of the dust, but a village unfriendly to Americans usually became an ally to the NVA or Viet Cong.[69]

Official policy did not allow anyone to fire his weapon

Two views of the security check point at bridge over Dak Oreno manned by the 173rd Airborne Brigade.
Photos courtesy of Bill Yates

at the Vietnamese unless fired upon.[70] So gunners normally kept their weapons on safe for fear of shooting an innocent Vietnamese who wore black pajamas like the Viet Cong guerrillas. Tim Wheat, however, expected trouble because his convoy had received sniper fire in that same area the day before, so when his gun truck crawled up the slight grade and rounded the bend to the right, he had his safety off. They were about a mile west of Phu Yen and two miles east of the check point and bridge over Dak Oreno when he saw seven khaki-clad Vietnamese (the uniform of North Vietnamese Army regulars) in the road. Ron Kendal called this in to 2LT Welton who told them to go ahead and fire. Tim clearly saw the expression of surprise of one Vietnamese looking at him. He then saw the muzzle of an AK47 sticking up behind one Vietnamese when he turned around and then he fired. The whole process took about 30 seconds and Tim never heard Ron say he had permission to fire.[71]

Ron Kendal saw them drop what he thought was a bundle of straw and instead of driving around it, he straddled it. Dye, driving the fuel tanker behind the gun truck, also noticed two black-clad Vietnamese peasants ahead walking alongside the road carrying rolled up bamboo. As the gun truck came alongside the peasants, Dye saw two Vietnamese toss the bamboo under gun truck, which exploded. They then jumped in the ditch beside the road. As soon as Wheat opened fire with his M-60 an explosion lifted the gun truck several feet in the air and blew out all eight rear duals. The blast threw Wheat back against the tailgate. He then picked up his M-14 and emptied a full magazine through the porthole at the Vietnamese alongside his gun truck.[72]

Welton generally rode in the jeep between the middle and rear of the convoy and was toward the back of convoy because a couple trucks had been hit between An Khe and Mang Giang Pass. He was half way up Mang Giang Pass when he received Kendal's call, "There's gooks in the road setting satchel charges." Welton told them to open fire and raced up the Pass.[73]

The convoy was running 15 minutes ahead of schedule and Kendal and Wheat believed they caught the NVA in the road by surprise, but the NVA concealed in the high ground on the passenger side of the road (north side) were waiting for them. There was an embankment about the height of the top of the gun truck armor on the right side of the road and the ground gradually sloped uphill across cleared field 30 to 40 yards to the tree line. The open area was covered by brush and trees bulldozed by the Army engineers into piles parallel to the road. The enemy had the high ground, which took away the advantage of the armor plating. After the gun truck opened fire then all hell broke loose. Kendal radioed back they were under attack and Welton told them to go ahead and return fire.[74]

Bobby Dye gunned his truck, drove around the gun truck and then was shot through both knees. An RPG hit Anderson's fuel truck mortally wounding him and setting it on fire. Dye drove two miles to the check point and the tank and paratroopers at the bridge over Dak Oreno had already heard the ambush. Dye stopped his truck, jumped out and fell, because he did not know he had been shot. Those vehicles not in the kill zone halted where they were and waited for the fight to end.[75]

Paul Marcum saw a flash in front of him and then shreds of rubber tires flying out. He then saw Dye's truck drive on past the gun truck. Anderson stopped his truck, got out and took cover on the downhill side of the road. Marcum stopped his truck 75 meters behind Anderson's on left hand side of road, also got out and then slammed the door to keep his dog in the cab. Then all hell broke loose and he knew they were in trouble. He then jumped down in the ditch in front of his truck while bullets pummeled his truck. Fuel was spilling out on the road and in the ditch. The enemy was so close he could hear them jabbering on top of the hill above the ten-foot embankment just across the road. Paul was firing his M-14 and when he came running up his weapon jammed. Bush's M-14 also jammed, so Paul took his friend's ammunition. They carried four to five 20-round magazines apiece. Paul sprayed bullets at the hill to keep enemy down and told Bush to watch behind them. Bush then saw two black-clad Vietnamese come around behind them. When Marcum turned around, they threw their hands up and he shot them. For years he wondered why they were there in the middle of the ambush and years later when Dye told him about seeing the two Vietnamese peasants toss the bundle of bamboo under the truck, he figured they were the same two.[76]

An 8th Group convoy halted behind them and their gun trucks came up and entered into kill zone. Mike Buirge was driving a truck in the rear of the convoy so he was on the other side of the Pass and did not even see the smoke from the burning tanker. The ¾-ton gun truck was in the back and 2LT Welton told them to stay back.[77]

After Marcum looked back around to the hill, the enemy tossed a lit torch that hit the tanker, bounced off and landed in the ditch, then caught the fuel on fire. The fire was spreading towards him and he knew he had to get to other side of road. He tossed a grenade, then ran across road with bullets hitting around him. He rolled over a metal gas line and busted up his knuckles. He thought he had been hit. He then watched for Bush to join him, but as the fuel in the tanker heated up, the four manholes on his tanker blew fire 50 feet in air. The heat scorched his neck. Looking back across the road, a grenade exploded and he thought they got Bush.[78]

From where the gun truck halted, it had a long field of fire along the convoy and tree line. The enemy was as close as the brush piles alongside the road and in the tree line. So Wheat burnt out the barrel of his M-60 with continuous spraying fire holding the enemy back from the tankers. Several tankers were hit by RPGs and set afire. The other gunner claimed his M-60 was jammed so he took cover in the right corner of the gun box and pointed out targets. Wheat finally convinced him to feed the ammunition belt to his gun. The barrel got so hot the bolt stripped the rim off a cartridge leaving the shell casing in the barrel. Wheat then pulled off the hot barrel with his bare hand and replaced it with the barrel from the other M-60. It did not leave a mark on his hand though. Meanwhile Kendal climbed into the gun box to provide covering fire with his M-14. After Wheat got the M-60 working, the other gunner fired an M-14. In the kill zone, the gun truck crew put up a fight against a hail of small arms, rockets and grenades from their disabled vehicle against an estimated 150 enemy soldiers. About ten minutes into the fight, Kendal said they were drawing fire from the left. Wheat lifted his M-60 off the mount and threw a heavy volume of fire down in that direction until that side quieted down. He then went back to suppressing the

This would have been the scene LT Welton would have seen coming into the kill zone from the rear. Enemy fire was coming from the high ground to the right. Photo courtesy of Billy Rumbo

The low ground to the right of the photo was where Wheat received some small arms fire.

Jacknifed fuel tanker. Photos courtesy of Billy Rumbo

enemy fire from the right side of the road. After a while, the bolt overheated and Wheat broke the two M-60s down and exchanged bolts while Kendal again laid down covering fire. Ron soon fired off all his magazines and asked what to do next. Wheat told him to start breaking down the M-60 belt and load rounds from it.[79]

Meanwhile, 2LT Welton raced up fast with his jeep and stopped inside the kill zone. He later remembered he had about eight tankers still in the kill zone and the rest behind him were not. One burning tanker had driven past the gun truck and stopped by a 30-foot embankment. The driver abandoned the truck. The 30-foot embankment concealed his truck from enemy observation and fire. Most of the drivers froze inside their trucks. Welton then stepped out of his jeep and ran along trucks with his M-79 grenade launcher while his gunner opened fire with his M-60. Some were burning when he arrived and others began burning after he reached them. If the trucks in the kill zone could drive he then instructed the drivers to drive on around the burning trucks to Pleiku and two or three did. Five tankers were on fire so Welton had to get the drivers out and have them return fire to keep enemy away from the trucks. While Welton talked to drivers on the passenger side, he only heard small arms and machine gun fire. He did not hear any RPG fire.[80]

Welton returned to his jeep to get ammunition for the gun truck when the gun trucks of the 8th Group convoy behind his asked permission to come up. The Lieutenant said they did not need them since they had the tank coming from the check point. One 8th Group gun truck entered the kill zone anyway and threw grenades. When they later found Anderson, he had small shrapnel holes in his head. Anderson may have been mortally wounded by friendly fire.[81]

From his position below the embankment, Paul Marcum kept looking back through the truck and saw something stand up, and then here came Bush. By then the smoke obscured his approach from the enemy and as soon as he arrived he said, "They got you, didn't they?" Paul replied, "I don't know." Then Bush said, "They got me." He was hit three times in the shoulder and arm. About that time the helicopter gunships arrived overhead and flew in so close Marcum could see the gunners. He thought they would hit them. A tank pulled up beside them, aimed its barrel up the

Vietnam War 43

road and fired. Bush wanted to get in the tank, but they laid there until the infantry finally arrived.[82]

About 15 to 20 minutes into the ambush, an M-60 tank rolled into the kill zone. An NVA soldier stepped around a tree to fire an RPG and the tank blew him in half. By then Tim had fired off all 8,000 rounds of M-60 ammunition, so he climbed out of his truck, ran over to the tank and borrowed another can of ammunition. By that time Kendal was totally exhausted from the fight. Welton also arrived with two cans of M-60 ammunition.[83]

A bunch of infantry hollered if anybody was wounded. Marcum answered, "Yes," and they then took Bush out. They collected dead enemy bodies and Paul went out to help. The tank had torn the body of the NVA soldier in two with the hind end resting on a stump. There was a machine gun laying there with roll of ammo. Paul came upon the enemy dead and they were young with cut clean jet black hair.[84]

After the ambush Welton loaded the drivers of the burning trucks in the other trucks. The remainder of the convoy had waited for what seemed hours, and then it just started moving. Welton led the convoy on to Pleiku without the gun truck. When Mike Buirge drove by the smoldering trucks, he saw half an NVA soldier on the road that had been hit by the M-60 tank. One fuel truck had jackknifed trying to get out of the kill zone.[85]

Upon inspection, the damaged gun truck had shrapnel in the gas tanks and the air lines were blown out so it had no brakes. Someone brought up four spare tires from An Khe to replace the duals on the gun truck. Kendal stuffed sticks in the fuel tank to stop the leaks and drove back to An Khe with a handful of damaged trucks. Tim rode back in another truck. The rest of the convoy continued to Pleiku. Kendal did not find out about the hole in the drive shaft until someone checked his truck out at An Khe. Upon return, Ron counted over 200 hits just on the driver's side door. He did not bother to count the number of hits on the right side. After the debriefing, Wheat was sent out on the ¾-ton the next day and then sent on R&R. Meanwhile, Ron Kendal rebuilt the gun truck and received permission to name it War Wagon.[86]

Dye's wounds were treated at the check point and he and the other wounded were medevaced to 71st EVAC. While at Pleiku, Welton visited Anderson in the hospital, but Anderson never regained consciousness and died ten days later.[87]

They returned to their barracks at An Khe that night. Paul took his flak jacket and clothes off to take a shower. He tossed his billfold on the cot and it flared open. He picked it up and found shrapnel halfway through it. The supply specialists returned to the 304th the end of that month.[88]

Lesson

Like the previous ambushes, the enemy initiated with explosives, but this time thrown under the trucks. It was also the first large scale ambush on a fuel convoy. It appeared as if the enemy laid an L-shaped kill zone but the base of the L was not ready when the convoy arrived. As usual, the enemy tried to destroy the lead gun truck first. The kill zone stretched over a mile in length and destroyed one gun truck and 10 tankers.[89] Again the gun trucks converged in the kill zone, increasing the fire power. This was becoming the pattern of 8th Group gun trucks, to rush into the kill zone and suppress the enemy until the nearest security force arrived.

Like the 4 December ambush, the crew of the gun truck realized they also needed to carry more ammunition. They added two more 4,000-round cans

Medevac site further up the road. Photo courtesy of Billy Rumbo

Anderson being treated for his wounds.
Photo courtesy of Billy Rumbo

of M-60 ammunition bringing the count up to 16,000 rounds, plus more M-14 magazines. They also added two spare M-60 barrels and carried an M-79 grenade launcher.

When Ron Kendal and Tim Wheat discussed the ambush, they agreed the driver needed to focus on the road and the gunners needed to focus on the surroundings. After Wheat left Vietnam in March, they added a third gunner to the gun box.

The location of the gun truck in the kill zone was fortunate as it allowed an 18-year old kid with an M-60 and 8,000 rounds of ammunition to lay down suppressive fire between the enemy and the trucks thus preventing greater disaster. The 359th realized it needed more gun trucks and built its second 2½-ton gun truck after this ambush.

Because the trucks usually spread out going up Mang Giang Pass, they regrouped above the Pass because they had encountered no enemy threat up there and it was close to bridge guarded by a tank platoon. The 1st Cav had previously secured the area, but moved north to I Corps Tactical Zone that January turning responsibility for the area over to the 173rd Airborne Brigade. A brigade then had responsibility for the same area previously defended by a division so it could not provide the same level of security. The convoy had evidently established a pattern with regrouping above the Pass, but never did that again.[90]

Welton felt there should not have been more than three trucks in kill zone. Because of regrouping, too many trucks were in the kill zone. Maintaining convoy interval of 100 meters would reduce the number of vehicles in the kill zone.[91]

In a guerrilla war the guerrillas often dress as civilians or even friendly troops. Senior commanders see the bigger picture of the war and have to make rules that protect the very people the war is fought to protect or defend. However, these rules as general as they are written do not always apply to every situation and the strict interpretation of them can put soldiers at risk. When Welton reported to his commander what happened, word came down from higher that they would either give him a medal or court-martial. He had instructed the gunner to fire even though he had not been fired upon. They were not even allowed to load a magazine in their weapons. 2LT Welton and Kendal were awarded the Bronze Star Medals with V device, Dye was awarded the Army Commendation Medal with V device, and Wheat received the Silver Star Medal.

The enemy attacks on convoys increased leading up to the commencement of the Tet Offensive on 31 January 1968, but failed to close down QL19. The truck drivers and convoy commanders learned a lot from the previous ambushes and the gun trucks clearly bought time for the convoy until the nearest security force could arrive and take over the fight. The convoys learned the importance of maintaining proper interval to limit the number of trucks in the kill zone. Those in the kill zone that could drive should clear the kill zone and drive to the safety of the nearest security check point. Those not in the kill zone should not drive into it. The gun truck crews also realized they needed more weapons and more ammunition. So while the convoys implemented these lessons over the next couple months, the enemy would continue to experiment with different ways to counter the suppressive fire power of the gun trucks.

Vietnam War 45

[15] Thomas Briggs email to Richard Killblane, July 16, 2009.
[16] LTC (R) Nicholas Collins telephone interview by Richard Killblane on 29 April and 18 June 2004.
[17] Colonel Joe O. Bellino, "8th Transportation Group; Sep 1967 – Sep 1968." n.d; and Interview with Phillip C. Brown and J.D. Calhoun by Richard Killblane at Ft Eustis, VA, 13 June 2003.
[18] Thomas Briggs email to Richard Killblane, July 16, 2009.
[19] Bellino, "8th Transportation;" Nichols interview; Walter Dan Medley interview by Richard Killblane at Pigeon Forge, TN, 6 August 2009. LTC Nick Collins, the S-3 during the ambush, remembered the claymore mine detonating next to SGT Collins' jeep and also remembered enemy detonated explosives in a culvert on the 19th vehicle, but Medley heard the mine was in a culvert and the tanker was the 19th vehicle in line of march, but the sketch map in the Bellino Report has it as the 12th vehicle.
[20] Bellino, "8th Transportation;" and Medley interview.
[21] Bellino, "8th Transportation;" Wolfe Interview; and Thomas Briggs, After Action Report to Commander C Company, 504th MP Battalion estimated 80 enemy.
[22] Medley interview; and evidence from photos taken of the damaged trucks.
[23] Brown and Calhoun interview.
[24] Medley heard it was a soldier in the 523rd named Jarmilio, but the Thomas Briggs claimed MP SGT Cox did the same thing.
[25] Briggs, After Action Report.
[26] Briggs, After Action Report; and Briggs email.
[27] Briggs, After Action Report.
[28] Burrell Welton telephone interview by Richard Killblane, 23 April 2013.
[29] Bellino, "8th Transportation."
[30] Briggs, After Action Report.
[31] The list of names came from the 512th TC Historical Supplement, 1967, and Memorial Service, 54th Transportation Battalion, Camp Addison, Republic of Vietnam, 6 September 1967.
[32] COL (R) John Burke telephone interview by Richard Killblane, 30 March 2004; and COL (R) Melvin M. Wolfe telephone interview by Richard E. Killblane, 31 March and 14 April 2004.
[33] Ibid.
[34] Bellino, "8th Transportation."
[35] Burke and Wolfe interviews.
[36] MAJ Nicholas H. Collins, "Battalion S-3 Notes," Headquarters, 54th Transportation Battalion, 5 March 1967; Burke interview, and Collins interview.
[37] Bellino, " 8th Transportation."
[38] Ibid.
[39] Burke interview.
[40] Bellino, "8th Transportation."
[41] Bellino, "8th Transportation;" and CPT Phillip T. Hall, Jr., 585th Transportation Company (Medium Truck Cargo) APO 96238, After Action Report, 11 November 1968.
[42] Ibid.
[43] Bellino, "8th Transportation;"and James Lyles, *Gun Trucks in Vietnam; Have Guns – Will Travel,* Wheaton, ILL: Rhame House Publishers Inc., 2012.
[44] Bellino, "8th Transportation."
[45] Ibid.
[46] Ibid.
[47] Ibid.
[48] Bellino, "8th Transportation;"and Lyles, Gun Trucks in Vietnam.
[49] Ibid.
[50] Bellino, "8th Transportation."
[51] Larry "Ted" Ballard interview by Richard Killblane, 6 August 2008.
[52] Ballard interview.
[53] Bellino, "8th Transportation;"and Lyles, Gun Trucks in Vietnam.
[54] Dennis Belcastro telephone interview by Richard Killblane, 12 April 2013.
[55] Ibid.
[56] Bellino, "8th Transportation;" and Dennis Belcastro claimed only 1LT Todd was the convoy commander, but the Unit History, 669th Transportation Company (Light Truck), 1 January 1967 – 31 December 1967 stated 1LT Richard J. Tillotson, 2nd Platoon Leader of the 669th Transportation Company, was the convoy commander, and was assisted by 1LT James R. Steel, 3rd Platoon Leader and 1LT Jerry L. Todd. The 1968 57th Unit History Report confirmed Todd was the assistant convoy commander.
[57] Belcastro telephone interview; and Dennis Belcastro, Narrative, "Living History Reenactment at Jamestown March 2000," Army Transportation Association Vietnam, http://grambo.us/belcastro/54guntruck.htm.
[58] Belcastro telephone interview.
[59] Belcastro telephone interview; and Belcastro, "Living History Reenactment."
[60] Bellino, "8th Transportation;" Belcastro interview; and 669th Transportation Company Unit History Report 1967.
[61] Bellino, "8th Transportation;" and Belcastro interview.
[62] Bellino, "8th Transportation;" and 669th Transportation Company Unit History Report 1967.
[63] 1LT Peter C. Thomas, Unit History Report, 57th Transportation Company, 1 January 1968 – 31 December 1968, 24 March 1969.
[64] Robert Dye email to Richard Killblane, December 9-11, 2009; Ron Kendal telephone interview by Richard Killblane, 15 April 2013; Timothy Wheat telephone interview by Richard Killblane, 16 April 2013; and Burrell Welton telephone interview by Richard Killblane,
[65] Michael Buirge telephone interview summary by Richard Killblane, 9 December 2009; Wheat interview; Kendal interview; and Welton interview.
[66] Dye email; and Award Citations for SP4 Ronald M. Kendall and SP4 Timothy Wheat, 7 February 1968.
[67] Paul Marcum telephone interview by Richard Killblane, 22 April 2013.
[68] Buirge interview, and Welton interview.
[69] Dye email.
[70] "Convoy personnel will not fire their weapons unless under attack. Rounds will not be chambered unless attack is imminent." COL Arthur B. Busbey, Jr., Subject: Hwy 19 SOP for Logistical and Security Elements, Headquarters I Field Force Vietnam, APO San Francisco 96350, 17 July 1968.
[71] Dye email, Kendal interview, and Wheat interview.
[72] Dye email, Kendal interview, and Wheat interview.
[73] Welton's interview.
[74] Dye email, Kendal interview, and Wheat interview.
[75] Buirge interview, Dye email.
[76] Marcum interview.
[77] Buirge interview.
[78] Marcum interview.
[79] Kendal interview, and Wheat interview.
[80] Welton interview, and Welton Bronze Star Medal Award General Orders Number 102.
[81] Welton interview.
[82] Marcum interview.
[83] Kendal interview, and Wheat interview.
[84] Marcum interview.
[85] Welton interview, and Buirge interview.
[86] Kendal interview, and Wheat interview.
[87] Gress email.
[88] Marcum interview.
[89] John Gress email to Richard Killblane, December 7, 2009.
[90] Welton interview.
[91] Welton interview.

Tet Offensive and After

The enemy attacks on convoys with mines, sniper fire and ambushes increased with frequency leading up to the commencement of the Tet Offensive on 31 January 1968, and through the next three months the enemy made a determined attempt to close down the flow of supplies to Pleiku along the QL19. Most of the attacks took place west of An Khe. In some cases the enemy planted as many as 30 mines in one night along that stretch of road. The truck drivers and convoy commanders learned a lot from the previous ambushes and the gun trucks clearly bought time for the convoy until the nearest security force could arrive and take over the fight. The convoys learned the importance of maintaining proper interval to limit the number of trucks in the kill zone. Those in the kill zone that could drive should clear the kill zone and drive to the safety of the nearest security check point. Those not in the kill zone should not drive into it. The gun truck crews also realized they needed more weapons and more ammunition. So while the convoys implemented these lessons over the next couple months, the enemy would continue to experiment with different ways to counter the suppressive fire power of the gun trucks.[92]

21 January 1968
54th Transportation Battalion

At approximately 0615 hours on 21 January 1968, a westbound convoy of 60 cargo trucks, four gun trucks and four gun jeeps under the control of the 54th Battalion departed Qui Nhon for Pleiku on QL 19. The 2½-ton gun trucks were still early designs with single steel walls and two M-60 machine guns, or Quad .50 gun trucks. A gun truck led each convoy march unit with gun trucks evenly spread throughout the convoy. Because of a shortage of radios on gun trucks, a gun jeep with a radio usually accompanied the gun truck. Each task vehicle only had one driver.[93]

As they had in the previous ambushes, again the enemy had placed mines in the road, and the convoy halted at Check Point 96 East for 30 minutes while EOD cleared the road. M48 tanks and M113 APCs manned the security check points along the route as the quick reaction force. At 1000 hours, approximately 500 yards east of Check Point 102, below Mang Giang Pass, the lead element of the convoy came upon a 5-ton tractor, which was attempting to hook up to a 5,000-gallon fuel trailer. Because this operation blocked the flow of traffic, the convoy commander drove forward and directed the clearance of the road. Once the road was open again, he then instructed his convoy to continue. By that time, the enemy opened with automatic and small arms fire on the driver's (south) side of the road. Gun trucks and gun jeeps immediately returned fire while the rest of the convoy continued to drive through the kill zone. Within five to ten minutes APCs from the road security element at Check Point 102 arrived and engaged the enemy. Tanks from Check Point 98 also arrived within ten minutes. Rear elements of the convoy approaching the area received between 40 and 50 rounds of automatic fire. Both APCs and tanks at the kill zone fired in the direction of the hostile fire.[94]

With only small arms fire this was more likely a hasty ambush taking advantage of the blocked traffic. Here the SOPs again worked. The gun trucks placed suppressive fire on the enemy so the convoy could clear the kill zone. Because the ambush did not stop any vehicles, the convoy drove through the kill zone as policy. The security force then responded to the kill zone but after the trucks had cleared it.

25 January 1968
54th Transportation Battalion

At 0600 hours on 25 January 1968, a westbound convoy under the control of the 54th Battalion consisting of 95 vehicles destined for Pleiku and 23 for An Khe departed the marshalling area at Cha Rang Valley. The 95 vehicles bound for Pleiku consisted of 65 M52 5-ton cargo trucks; 19 M35 2½-ton cargo trucks; five 2½-ton gun trucks, radio-equipped radio gun jeeps and two 5-ton maintenance trucks. A gun truck led each convoy march serial with gun trucks accompanied by radio-equipped gun jeeps evenly spread throughout the convoy. Each task vehicle only had one driver. M48 tanks and M113 APCs manned the security check points along the route as the quick reaction force.[95]

At approximately, 1015 hours, the convoy received automatic and small arms fire from both sides of the

Vietnam War   47

road about 12 kilometers west of Mang Giang Pass. The gun trucks and convoy personnel returned fire and within 10 minutes elements of the 2/1st Cavalry arrived with APCs and tanks. This ended the enemy fire for a while; and after all the firing stopped, the convoy proceeded west for approximately 500 yards when the convoy again came under fire. The two kill zones spanned 1,000 meters. The NVA opened fire on the second and third vehicles in the convoy with rockets, heavy machine guns, grenades and small arms fire from both sides of the road. A machine gun position was later discovered approximately 25 yards from the passenger (north) side of the road. One 2½-ton gun truck and one 2½-ton cargo truck were damaged, and a Pacific Architects & Engineers (PAE) commercial tractor was destroyed. Two drivers were killed and one wounded. The reaction force immediately left their previous location and raced to the new kill zone where the fire again ended. Three armed helicopters arrived at approximately 1045 hours. Two medevac helicopters arrived within 10 minutes after receiving the request. Two officers, members of the engineer team, were wounded, one fatally, while clearing explosive ordnance from the site.[96]

The enemy had changed its tactics again. Evidently the enemy had studied the gun trucks and noticed they stayed in the kill zone while the task vehicles escaped. Consequently, the enemy planned two kill zones possibly to see how the convoy and security force would react. The enemy may have planned the first ambush as a decoy to draw off the gun trucks hoping that the unprotected trucks escaping the first kill zone would drive into the main kill zone. In this case, the gun trucks defended the halted convoy long enough for the reaction force to arrive and then left. This placed the reaction force only a short distance behind the convoy when it drove through the next kill zone. This allowed the armored cavalry to respond rapidly and turn the fire back on the enemy. At this time the enemy feared the fire power of the tanks and APCs over the gun trucks.

The enemy watches for patterns. So varying procedures can keep the enemy off balance. Some gun trucks should remain with the vehicles that escape the kill zone and having a combat arms force following a close distance behind but out of sight of the convoy provides a great tactical surprise for the enemy that ambushes the convoy.

30 January 1968
54th Transportation Battalion

On 30 January 1968, a westbound convoy under the control of the 54th Battalion departed for Pleiku on QL19 at approximately 0600 hours that morning. The convoy consisted of 80 cargo trucks, seven gun trucks, eight gun jeeps and three Quad .50s. This exceeded the 1:10 ratio of gun trucks to prime movers.[97]

Upon arrival in An Khe, the convoy was joined by three APCs and one tank from the security force of the 173rd Airborne Brigade. The additional security element was dispersed toward the front of the convoy. Since the convoy was about to pass out of the area of operation of the 173rd, the additional security element pulled out of the convoy and stopped at Check Point 102. Approximately one mile west of Check Point 102, the convoy came under enemy fire initially from mortars followed by small arms and automatic fire from a platoon-size enemy force. The convoy personnel immediately returned fire. In addition, the 173rd Airborne Brigade security element advanced from Check Point 102 and an element for the 4th Infantry Division security element moved west to engage the enemy. F111As, F104s and helicopter gunships made air strikes. Two US personnel were slightly injured; one 5-ton tractor and reefer (refrigeration van) were damaged. No enemy dead or wounded was found.[98]

This time the NVA initiated with mortars, which proved inaccurate for initiating an ambush. In this case the role of the security force was much more proactive. Their presence along with the high number of gun trucks suppressed the enemy attack minimizing any damage to vehicles or casualties.

31 January 1968
124th Transportation Battalion

1 February was the Lunar New Year or Tet, one of the most popular holidays in Vietnam. It was like Christmas and New Years rolled into one holiday. The insurgents had agreed to a ceasefire so that the Army of the Republic of Vietnam (ARVN) soldiers could go home for the holidays. On 31 January 1968, the enemy launched a nationwide offensive.[99]

That afternoon, 1LT David R. Wilson, 64th Medium Truck Company, led an eastbound convoy returning from Pleiku on QL 19. The convoy consisted of twenty-four 5-ton tractor-trailer combinations, two radio-equipped gun jeeps and three 2½-ton gun trucks from the 124th Battalion. Seven 5-ton cargo trucks, with a lead 2½-ton gun truck and trail radio equipped gun jeep from the 54th Battalion made up the rear of the convoy. Two Quad .50 gun trucks attached from B Battery, 4th Battalion, 60th Field Artillery accompanied the convoy. The loads consisted of four 5,000-gallon fuel tankers, two Class II (individual equipment) and IV (construction material) loads, five Class IV loads and eight Class V (ammunition) loads. The trucks from the 54th Battalion hauled engineer Class IV loads. Wilson employed a convoy line up of gun jeeps front and rear with gun trucks evenly spaced throughout the convoy. He placed the Quad .50s in the middle of the convoy. At that time convoy commanders rode in the front of the convoy either in front of or behind the lead gun truck.[100]

The convoy made it safely to An Khe and departed for Pleiku at 1430 hours and reached the base of the Mang Giang Pass. Where the road grade started to raise slightly, the area was cleared for approximately 100 yards on both sides of the road. This was the stretch of road where the French Mobile Group 100 was destroyed and the site of recent ambushes became known as "Ambush Alley." At 1520 hours, the lead vehicles started up the Pass when the enemy hidden the wood line initiated the ambush with mortars on the middle of the convoy, in the vicinity of Check Point 102.[101]

The lead gun truck escorted the lead elements of the convoy to the top of the Pass in accordance with the SOP. However, many of the vehicles behind it had halted in the kill zone and were subject to an intense enemy mortar and small arms fire. Upon hearing the firing behind him, 1LT Wilson, immediately ordered his driver, SP4 Brammer, to turn the jeep around and reenter the kill zone to make an estimate of the situation and so he could insure the safe passage of the rear element of the convoy. Wilson's gunner, SP4 Earnest W. Fowlke, blazed away with the M-60. Upon reaching the edge of the kill zone, he saw that the enemy fired small arms, automatic weapons, 60mm and 82mm mortars into a 400 yard long kill zone. Wilson then turned the jeep around and headed back up the Pass. The jeep had traveled 50 meters when a mortar round hit it causing it to burst into flames, killing Wilson instantly and mortally wounding Fowlke. Brammer was only slightly wounded and the burning jeep rolled into a ditch.[102]

The lead gun truck escorted the convoy safely to the top of the Pass and then returned to the kill zone to escort the rest of the convoy out. As it came down the Pass, the two machine gunners, SP4 Howell and SP4 Bushong, saw the two Quad .50s racing up the Pass to safety. One driver said they only saw one return fire as it beat other trucks out of the kill zone. The 2½-ton gun truck stopped at the edge of the kill zone 100 meters east of the Pass. The middle gun truck raced up to a position about 50 meters from the first. To obtain better fields of fire and maximum coverage, the gun trucks remained in the kill zone.[103]

The last gun truck of the 124th Battalion headed up the road but was stopped by MPs at Check Point 102. Evidently unaware of the ambush the driver, SP5 Jimmie Jackson, jumped out to find out why the convoy had stopped. None of the gun trucks had radios. Then the rear gun jeep arrived with SGT Welch, PFC Jimmy Tidwell and PFC Cansans. PFC Tidwell informed Jackson and the security force that the convoy up ahead was ambushed. Tidwell jumped into the driver's seat of the gun truck and Jackson climbed in the gun box to man a machine gun. They drove up to Wilson's burning jeep and returned fire on the enemy. They remained in position until the MPs instructed the convoy to move forward. Jackson and SP4 Green were wounded in the fight. Only Wilson's jeep was destroyed. There were two APCs and one tank in the area. Helicopter support did not arrive for 20 minutes.[104]

The enemy was experimenting with mortars in a convoy ambush and this time launched the attack on the middle of the convoy instead of the front as it had in the past. This allowed the convoy commander and lead gun truck to escape. However, at great risk to life, 1LT Wilson and his crew drove back into the kill zone to save the lives of their drivers. For their personal bravery and sacrifice, 1LT Wilson and SP4 Fowlke were posthumously awarded the Silver Star Medals. The 124th Battalion named their camp after Wilson. While extremely brave, driving into a kill zone in an unarmored vehicle was dangerous.

1LT John M. Arbuthnot II wrote the following as lessons:

Quick reaction is required of all individuals in determining the type of attack and the limits of the kill zone. When small arms fire is combined with a mortar attack the driver must do more than return heavy volume of fire. Personnel must get out of the kill zone, preferably, and/or obtain overhead cover. Air cover should be immediately available to locate the mortar positions and then silence these positions or request artillery support. It is highly desirable for the aircraft to provide continuous overhead cover for the convoy and not just be on call. The five or ten minutes lost waiting for the aircraft to arrive often means the difference between a few casualties or a large number and minor or major damage to vehicles and equipment.

Since an ambush is generally of limited duration, drivers must be constantly alert and capable of reacting instantly with a heavy volume of effective fire when fired upon. Simultaneously the personel must move quickly through the kill zone and direct maximum fire at the most logical vantage points. The convoy commander must immediately identify his location, notify the road security forces, request air and/or artillery support and proceed to facilitate movement of his convoy through the kill zone.[105]

By attacking the middle of the convoy, this split the convoy in half. The lead vehicles drove on to safety while the vehicles behind the kill zone halted awaiting a decision. As the enemy increasingly began to attack the middle or rear of the convoys, convoy commanders realized the front was not the place to be if they wanted to influence the action. The key decisions had to be made either in or behind the kill zone. For that reason, convoy commanders would learn to ride in the rear of the convoys.

The lack of radios in the gun trucks also caused confusion in the rear when the MPs stopped the last gun truck. Without knowing what was happening ahead of them and with no communication with the convoy commander, the last gun truck was unable to respond in a timely manner.

Everyone needed to know what to do in an ambush, especially attachments. Besides being critical of the performance of the artillerymen in the Quad .50s, the truck drivers were disappointed with the delayed arrival of the helicopters. The machine guns on the gun trucks were effective weapons against other direct fire weapons but not against mortars hidden in the woods. Helicopters could more effectively spot the mortars which either they or artillery could then destroy. Similarly the MPs at the check point only knew to prevent vehicles from entering the kill zone. While it may have seemed obvious that the gun trucks should have been allowed through, fear causes a form of paralysis and soldiers do only what they know.

As 1LT Arbuthnot clearly realized, if air cover ended the ambush, its presence would most likely prevent it; but unfortunately it would take a few more years before air cover became required for convoys.

7 February 1968
54th Transportation Battalion

On 7 February 1968, a westbound convoy on QL19 of 67 cargo trucks, six gun trucks, four gun jeeps and one maintenance truck under the control of the 54th Battalion departed Cha Rang Valley at 0630 hours. Each 2½-ton gun truck escorted a march serial of around ten task vehicles and the maintenance truck brought up the rear. Each gun truck had at least two M-60 machine guns and possibly one M-2 .50 caliber machine gun. The gun jeeps were armed with M-60s and radios with one gun jeep up front and another in the rear.

At approximately 1010 hours, after passing Check Point 92, the convoy came under small arms and automatic fire from 50 to 60 enemy soldiers hidden in the tree line south of the road. The enemy fired two rockets at the convoy from a mound halfway between the road and the tree line. One rocket hit a 5-ton cargo truck carrying a load of mortar ammunition, which was destroyed. The exploding rounds also damaged the next vehicle in the convoy. The enemy force then began to advance from the tree line but was driven back by the fire power of the gun trucks. The gun trucks, which had cleared the 200 meter long kill zone, returned to fire on the enemy positions. Within 15 minutes, six to eight APCs and two to three tanks arrived on the scene to engage the enemy. The helicopter gunships arrived within 15 to 20 minutes.[106]

The enemy laid a linear kill zone and initiated

with rockets. The Americans only had four men slightly wounded, but the enemy lost six killed and one wounded. One truck was destroyed by the rocket, the next 5-ton cargo truck had its gas tank damaged by the exploding mortar rounds and many of the vehicles including the gun trucks received flat tires from enemy fire. The gun trucks provided immediate security until the tanks and APCs arrived to take over the fight, and then the convoy continued on to Pleiku.[107]

In February, the M-16s and M-2 .50 caliber machine guns began to arrive. The three companies attached to the 54th Battalion traded in their M-14 rifles for M-16s and conducted training on handling, cleaning and firing from 9 to 21 February. The battalion trained the unit armorers on M-16 maintenance from 12 to 14 February. Similarly the battalion conducted a class on the M-2 .50 caliber from 21 to 24 February.[108]

13 February 1968
124th Transportation Battalion

On 13 February, a westbound convoy on QL19 of 28 task vehicles, one maintenance vehicle, two gun jeeps, three 2½-ton gun trucks and three Quad .50s under the control of the 124th Battalion departed An Khe at 1335 hours for a return trip to Pleiku. This convoy had a gun truck ratio of 1:5 just counting the gun trucks. By policy, a gun truck would lead the convoy followed by a radio-equipped gun jeep and likewise another pair would bring up the rear with the remaining gun truck and Quad .50s evenly distributed throughout the convoy.

Around 1500 hours, 200 yards west of the base of the Mang Giang Pass, the enemy ambushed the lead elements of the convoy with mortars and small arms. The convoy personnel and gun trucks immediately returned fire. The convoy commander urged one of the Quad .50s from the 4th Battalion, 60 Field Artillery to fire on the suspected enemy mortar position while the convoy drove through the kill zone. The security forces of the 173rd Airborne Division arrived and attacked the enemy within 10 minutes. Air strikes were conducted against the suspected mortar positions. The enemy had only fired seven mortar rounds and caused no vehicle damage or wounded any drivers.[109]

The enemy seemed to experiment with different types of casualty producing weapons to initiate the ambush with and seemed focused on mortars. Again mortars proved an inaccurate weapon to initiate the ambush. So with the mortars not stopping the convoy, the Quad .50 laid down suppressive fire while the rest of the convoy continued through the kill zone.

21 February 1968
27th Transportation Battalion

At approximately 0715 hours, 21 February, a westbound convoy on QL19 under the control of the 27th Battalion departed the marshalling yard at Cha Rang Valley for Pleiku. The convoy consisted of 54 task vehicles, four 2½-ton gun trucks, four gun jeeps and a Quad .50 evenly spaced throughout the convoy. So the five gun trucks gave the convoy the desired 1:10 gun truck ratio. There were just about enough radio-equipped gun jeeps for each gun truck. The Quad .50s were usually placed somewhere in the middle of the convoy.[110]

At approximately 0950 hours, the convoy came under fire from automatic and small arms fire and B-40 rockets in approximately 300 to 400 meter kill zone in Ambush Alley between Check Points 89 and 96. Because the convoy maintained the proper 100-meter interval, only three cargo trucks, two from the 58th and one from the 444th Truck Companies, were in the kill zone. The convoy personnel returned fire in the direction of an estimated 10 to 12 NVA south of the highway. The Quad .50 moved into the small kill zone and was credited with killing at least one NVA soldier. APCs from the 173rd Airborne Brigade arrived approximately five to ten minutes after the call and engaged the enemy force. The tactical force also called in artillery. Three vehicles including a task vehicle and the Quad .50 were damaged and three personnel were wounded. One enemy was killed and another wounded was recovered along with the discovery of numerous foxholes.[111]

This was a linear ambush and the enemy was setting up smaller kill zones. Consequently, the 100-meter interval limited only three task vehicles and one gun truck in the 300 to 400 meter long kill zone, thus reducing the amount of damage to vehicles.

4 March 1968
54th Transportation Battalion

On 4 March, a westbound convoy under the control of the 54th Battalion departed Cha Rang Valley for Pleiku at approximately 0600 hours. The convoy consisted of 104 task vehicles, eight gun trucks and four gun jeeps. The gun truck ratio without the gun jeeps was 1:14, but the limited fire power of the additional four gun jeeps brought the gun truck ratio down below 1:10. As standard procedure, a gun truck would lead the convoy followed by a radio-equipped gun jeep and a similar pair in the rear of the convoy. The remaining four gun trucks would be evenly spaced throughout the convoy.[112]

At approximately 0900 hours, the convoy was held up at Check Point 89 by the tactical security force due to enemy activity ahead in Mang Giang Pass. The convoy was allowed to proceed at approximately 1130 hours with the escort of one tank and two APCs from the 173rd Airborne Brigade. At approximately 1145 hours, the enemy initiated the attack with three mortar rounds followed by 20 minutes of heavy small arms and automatic fire. Convoy security immediately opened fire in the direction of the enemy which was well entrenched in the tree line on the north side of the road. The convoy also received sporadic fire from the south side of the road. The enemy force was estimated at about 50 personnel. Two Quad .50s from the 4th Battalion, 60th Artillery, which were traveling with the convoy and one from the 27th Transportation Battalion convoy serial, which followed behind the 54th convoy, fired upon the enemy positions throughout the kill zone, estimated to be between 500 and a 1,000 meter long. A reaction force of one tank, four APCs and four gunships arrived within five minutes.[113]

Eight convoy personnel were wounded, three from the artillery unit. One wounded soldier died on 6 March, from wounds received in the battle. One vehicle was destroyed and five vehicles and two trailers were damaged. The convoy remained in place on the highway until 1430, at which time they turned around under the escort of MPs and returned to An Khe.[114]

8 March 1968
54th Transportation Battalion

On 8 March, another 54th Battalion westbound convoy had departed from Cha Rang Valley on QL19 for Pleiku at approximately 0600 hours. The convoy consisted of 79 task vehicles, four gun jeeps and five gun trucks. This convoy did not have the required 1:10 ratio of gun trucks to task vehicles unless one counted the gun jeeps.[115]

At approximately 0830 hours, the third gun truck of the first serial was hit with a captured claymore mine that destroyed the vehicle. The explosion was followed by heavy small arms and automatic fire from both sides of the road. Three Quad .50s from the 4th Battalion, 60th Artillery traveling with the convoy, joined by a company of tanks and APCs of the 173rd Airborne Brigade, which was in the area, engaged the enemy. The company size NVA force attempted to repel the flanking action of the tactical security force but was driven back after 15 minutes of heavy contact. The enemy had destroyed one gun truck and four cargo trucks in that kill zone. The first ambush wounded six truck drivers and two soldiers from another company and killed the company commander of the 173rd. The convoy reformed and was allowed to proceed after a 20-minute delay.[116]

At approximately 0915 hours, two kilometers west of Check Point 102 in Mang Giang Pass and ten miles past the first kill zone, a task vehicle in the first serial hit a mine then small arms fire hit the cab of the disabled vehicle wounding the driver. B-40 rockets then ignited the JP4 that the 5-ton cargo truck hauled. A reinforced enemy platoon opened fire with small arms, automatic fire and rockets on the convoy in a kill zone of approximately 300-500 meters in length. The convoy security element fired in the direction of the enemy positions as the convoy maneuvered around the burning vehicle. Tactical security forces from the 173rd and the 4th ID arrived within five minutes and engaged the enemy. Only four truck drivers were wounded with two trucks damaged and another destroyed. One MP company commander was killed in this second kill zone. No enemy dead or wounded was recovered.[117]

Again the enemy employed two kill zones. They also initiated the ambush on the gun trucks. The casualties were inflicted by the initiation of the ambush but the quick reaction by the gun trucks and security force reduced further casualties. Because of the additional fire power of the 173rd Airborne escort, the convoy continued to drive through the kill zone.

16 March 1968
124th Transportation Battalion

On 16 March 1968, a convoy of 17 cargo trucks, three gun trucks, one gun jeep and a ¾-ton maintenance truck with an M-60 machine gun, under the control of 124th Battalion, departed Dak To for Pleiku on QL14 at approximately 1430 hours. 1LT Jimmie W. Reed, of the 64th Medium Truck Company, was the convoy commander. MPs from the 4th Infantry Division (Mechanized) escorted the convoy. The 124th Battalion pushed convoys of 2½-ton and 5-ton cargo trucks to the smaller base camps along the Cambodian border. QL14 cut through dense jungle. The jungle provided adequate concealment for the enemy.[118]

At approximately 1800 hours the rear element of the convoy came under heavy small arms and automatic fire from a platoon size enemy force 12 miles south of Kontum. The convoy personnel immediately returned fire in the direction of the enemy and moved rapidly out of the estimated 100 to 150 meter long kill zone. Tanks from the nearest tactical force arrived within three to five minutes and gunships arrived within five to ten minutes. Only the maintenance vehicle in the rear was damaged and the machine gunner, SP4 Robert W. Hardesty of the 64th, was killed.[119]

As a result of the ambush, the 64th recommended the maintenance gun truck have more protection against small arms and the crew came up with the idea of slanting the armor so bullets would glance off. 8th Group approved the design. They also replaced the M-60 with a .50 caliber machine gun for greater range and firepower.[120]

23 March 1968
Night Shuttle

While the day drivers slept, the night shuttle drivers drove empty trucks 8 miles to the supply yards at Qui Nhon to pick up loads and then drove the trucks back to the marshalling yard where they were staged for the convoys the next morning. On 23 March, a night shuttle convoy from the port of Qui Nhon was proceeding west on Highway 1 toward loading sites in Cha Rang Valley. Highway 1 was heavily trafficked by civilian traffic during the day and normally considered safe. At approximately 0015 hours, the convoy consisting of five task vehicles, one gun truck and one gun jeep, approached the bridge guarded by the Koreans. The convoy commander, 1LT Paul J. Stegmayer, of 2nd Medium Truck, observed a pipe line fire in the vicinity of Tuy Phovc. After reporting the same, 1LT Stegmayer proceeded with his column. As the convoy reached the site of the fire, an explosion occurred on the north side of the road near 1LT Stegmayer's jeep, followed by heavy small arms and automatic weapons fire. Although both 1LT Stegmayer and his driver received wounds from flying glass and shrapnel, they were able to cross over the bridge at the site of the pipe line fire. Due to the intense enemy fire, only the jeep and one task vehicle were able to clear the kill zone. Despite great personal danger, 1LT Stegmayer, braving a withering hail of bullets, crossed back over to bridge on foot to take control of the drivers and insure that they could clear the scene. Moving from vehicle to vehicle, Stegmayer assured himself that all drivers were out of their vehicles and had taken up positions to engage the enemy. He crossed back to his jeep to radio reports to Battalion and adjust illuminating artillery rounds.[121]

With arrival of a reaction force of three gun trucks, one gun jeep and a Quad .50, 1LT Stegmayer again crossed over the bridge to direct flanking fire into the suspected enemy positions. The enemy force estimated at 15 broke contact and fled the area. All six vehicles in the convoy received small arms and automatic weapons fire. Four personnel were wounded.[122]

Intelligence reports indicated that the enemy's mission was to destroy the dual bridges (railroad and highway) at the site of the pipe line fire thus cutting a vital link on the only main highway between Qui Nhon and major tactical forces to the north and west. This was probably not a planned ambush. With the arrival of the shuttle convoy, the enemy, for reasons unknown, fired on the column. It has been recommended that the enemy may have mistaken the convoy as a reaction force investigating the pipe line fire. The action by 1LT Stegmayer and his men contributed to the failure of the enemy to accomplish their mission of interdiction of lines of communication to the north and west.

2 April 1968
54th Transportation Battalion

While most ambushes on QL19 occurred west of An Khe, the enemy conducted a small ambush against a convoy of the 54th Battalion in the lowlands area east of An Khe Pass on 2 April. The Korean Tiger Division patrolled the rice paddies below An Khe Pass and up until this time, the enemy only planted mines and sniped on passing convoys. A reinforced squad ambushed the convoy with intensive automatic fire wounding four truck drivers and damaging seven trucks. The gun trucks forced the enemy to withdraw after only eight minutes of fighting.[123]

Like the 2 September 1967 ambush, this ambush may have been a dress rehearsal to test the success of ambushing in this area. The Koreans would brutally punish the locals in the area and as one driver remembered, "There was a My Lai massacre in the Korean sector every week." So the drivers felt relatively safe in the lowlands.

End of the Tet Offensive

The enemy's Tet Offensive had ran its course and turned into a bitter military defeat. The Viet Cong had been all but wiped out and all that was left to fight was the NVA. The NVA retreated back across the border to their sanctuaries to refit, rebuild and plan for their next offensive. Consequently, the 8th Transportation Group had won its campaign to keep the supply line open. With the end of the Tet Offensive, ambushes on convoys also died down. The NVA would launch a couple more offensives in May and August but with none of the violence of the Tet Offensive. As the enemy took more time to plan, convoy ambushes would become less frequent but in some cases more deadly. The randomness of these ambushes indicated that the enemy only hoped to inflict casualties to wear down American moral and will to continue the war.

Kill zones were becoming smaller concentrating fire into approximately 300 meters. The enemy had tried every type of weapon system but found mines and RPGs the most lethal to initiate an ambush. Likewise, 8th Group learned overwhelming firepower won the fight and all gun trucks were racing into the kill zone. Although they could not see the muzzle flash of enemy weapons during the day, the gunners had learned to fire in every direction. This caused them to expend ammunition fast so they began to fill the bed of the gun truck with as many ammunition cans as it would hold. The fight usually ended when either the air support or ground support arrived, which the truck drivers considered was too slow. They wanted the air support to accompany the convoys rather than wait for the call.

12 May 1968
27th and 54th Transportation Battalion

QL1, the coastal highway, was heavily trafficked by civilians as well as military and aside from the attack on the night shuttle run, daytime traffic on the coastal highway seemed relatively safe. At 1000 hours on 12 May 1968, two convoys approached Bridge 329 (vicinity BR 985856) along the coastal highway. The southbound convoy of 17 task vehicles under the control of the 27th Battalion came under M-79 and automatic fire 200 meters south of the bridge from a squad size enemy force hidden in a tree line 150 to 200 meters on the west side of the road. A northbound convoy of 31 task vehicles plus security, under the control of the 54th Battalion was also approaching the same location when the fire began. Both convoys increased speed to clear the 200 meter long kill zone.[124]

The 27th Battalion convoy managed to clear the

kill zone with only minor damage to the vehicles but the 54th Battalion convoy suffered one driver killed in action and the 240th Quartermaster Battalion had one POL tanker driver wounded. The driver of a 2½-ton cargo truck was killed and ran off the road onto a small bank. A gun jeep and two gun trucks rendered immediate assistance while directing the convoy through the kill zone. They evacuated the casualties and the convoy continued north without further incident.[125]

The 27th Battalion convoy continued south and reached Bridge 376 (BR 930779) at approximately 1010 hours. There it came under automatic and M-79 fire from a platoon-size enemy force 200 meters from the east side of the road. Upon receiving enemy fire the lead gun truck pulled over and engaged the enemy while the convoy raced through the kill zone. Only one driver was wounded and four trucks damaged with bullet holes from both ambushes.[126]

Lesson

The enemy again set up two kill zones, both on a heavily trafficked highway that convoys did not expect any trouble. Speed was security in both cases, so the vehicles not in the kill zone raced through it. The gun trucks and gun jeeps concentrated around the 2½-ton truck and medevaced the driver. The role of gun trucks was to not leave any drivers in the kill zone, and because of the threat of a second kill zone, at least one gun truck needed to escort the remainder of the convoy out of the kill zone.

14 August 1968
54th Transportation Battalion

At approximately 1215 hours, 14 August 1968, a convoy under the control of 54th Battalion departed Cha Rang Valley on QL19 for a line haul trip to Pleiku. The convoy consisted of 68 task vehicles, seven gun trucks, five gun jeeps armed with M-60 machine guns, and one Quad .50 gun truck.[127] This convoy had well over the desired 1:10 ration of gun truck to task vehicles even without counting the gun jeeps. The gun trucks in the kill zone came from the 512th and 523rd Transportation Companies. The 523rd had constructed five M54 5-ton gun trucks with steel plate reinforced with sandbags. Each crew consisted of a driver, grenadier with the M-79, and two machine gunners.[128] The 512th had authorized the construction of five M35 2½-ton armor-plated gun trucks with crews of two to five personnel.[129] By then the drivers had M-16 rifles.

At 1545 hours, at the first serial of the convoy proceeded west past an area approximately two miles west of Bridge 34, an enemy force dressed in ARVN Marine uniforms attacked the convoy with small arms and B-40 rocket fire. The enemy force was estimated at between a platoon and a company. The first ambush opened fire with a 250 to 300 meter kill zone. Four gun trucks, one Quad .50 gun truck and one gun jeep immediately returned fire then rolled into an 800 to 1,000 meter kill zone a couple kilometers down the road. A reaction force of six APCs and three helicopter gunships arrived within five minutes after contact. All task vehicles cleared the kill zone, but four of those vehicles suffered minor damage and one was heavily damaged. The convoy had four men wounded and one driver from the 1/69th Armor was killed. The convoy commander reported 12 enemy troops hit by return fire. After the security forces swept the area of contact, they discovered four enemy dead. The five wounded US Soldiers were medevaced to the 71st Medical Evacuation Hospital. Of those, two were treated and released to duty.[130]

Lesson

The enemy again varied his tactics by disguising his forces as friendly ARVN troops. Convoy personnel should be suspicious of everyone and everything. Be aware of anything unfamiliar and look for clues. While the enemy may be able to dress like friendly soldiers, they make mistakes in dress and behavior. The NVA also employed two kill zones, one a diversionary possibly an attempt to draw off the gun trucks while the secondary kill zone caught the rest of the convoy defenseless.

Big Kahuna - Because of the shortage of armor kits, 8th Transportation Group truck companies experimented with sliding M113 Armored Personnel Carrier (APC) hulls on the beds of 5-ton cargo trucks. The APC mounted gun truck was born out of the necessity of finding a way to replace the armor destroyed during a period in which there was a shortage of steel plating in the supply system. Big Kahuna was armed with two M-60 machine guns, one .50 caliber machine gun and one M-79 grenade launcher.

Gambler - Typical 8th Group 2½-ton gun truck with armored plating. This vehicle had sandbagged floors, 2 pedestal-mounted M-60 machine guns with sandbagged emplacements for added protection, and a M-79 grenadier. The vehicle commander controled the vehicle from his ring-mounted .50 cal machine gun. Photos courtesy of the US Army Transportation

The Hawk - Another version of the versatile gun truck was this M-54A2, 5-ton truck with armored plated cargo bed and cab. The floors and walls had additional sandbag protection. This vehicle was mounted with two M-60 machine guns, a M-79 grenade launcher and .50 cal machine gun. Also, the crew was armed with M-16 rifles and a M-79 grenade launcher. Photo courtesty of the US Army Transportation Museum

---

[92] LTC Jack C. Utley, Operational Report – Lessons learned, Headquarters, 54th Transportation Battalion (Trk), Period Ending 30 April 1968, 17 July 1968.
[93] Bellino, "8th Transportation."
[94] Ibid.
[95] Ibid.
[96] Ibid.
[97] Ibid.
[98] Ibid.
[99] Ibid.
[100] Ibid.
[101] Ibid.
[102] Ibid.
[103] Ibid.
[104] Ibid.
[105] Ibid.
[106] Bellino, "8th Transportation;" and Utley, Operational Report.
[107] Bellino, "8th Transportation;" and Utley, Operational Report.
[108] Utley, Operational Report.
[109] Ibid.
[110] Ibid.
[111] Ibid.
[112] Ibid.
[113] Bellino, "8th Transportation;" and Utley, Operational Report.
[114] Ibid.
[115] Bellino, "8th Transportation."
[116] Bellino, "8th Transportation;" and Utley, Operational Report.
[117] Bellino, "8th Transportation," reported one truck damaged and one destroyed, with one soldier wounded in the second kill zone. Utley, Operational Report claimed two trucks damaged and one destroyed, and four truck drivers wounded and one MP company commander killed in the second ambush.
[118] Bellino, "8th Transportation."
[119] Ibid.
[120] 1LT William J. Wilkins, Unit History Report, 64th Transportation Company (Medium Truck), 1 January 1968 – 31 December 1968, 7 April 1969.
[121] Bellino, "8th Transportation."
[122] Ibid.
[122] Utley, Operational Report.
[123] Bellino, "8th Transportation."
[124] Ibid.
[125] Ibid.
[126] Ibid.
[127] Ibid.
[128] 1LT Harris T. Johnson III, Unit History, 523rd Transportation Company (LT TRK), 1 January to 31 December 1968, 31 January 1969.
[129] CPT Paul Forster, Unit History, Annual Supplement, 512th Transportation Company (LT TRK), 1 January to 31 December 1967, 18 March 1968.
[130] Bellino, "8th Transportation."

88th Transportation Company, 124th Transportation Battalion

The 88th Transportation Company had arrived in Vietnam in August 1966. It was originally a light truck company and was the first truck company that located at An Khe. It would push cargo out to the smaller camps along the Cambodian border. Its runs usually began with pushing to Pleiku in the morning then going to Qui Nhon in the afternoon to pick up another load. Because it was the only line haul truck company at An Khe, all the trucks in its convoys belonged to its company its convoys were generally smaller than those originating from Qui Nhon. When the 124th Battalion arrived in the summer of 1967, the 88th Medium Truck fell under its control. After the September ambush, CPT Terrance J. McGinty personally supervised marksmanship training for his drivers to include one-handed firing positions from the cab of trucks. In April 1968, the 88th traded in its 5-ton cargo trucks for 12-ton tractors and trailers and became a medium truck company.[131]

23 August 1968

At 0815 hours on 23 August 1968, three gun trucks and two gun jeeps of the 88th Transportation Company escorted a convoy of 30 trucks from An Khe to Pleiku along QL19. Many of the trucks were commercial contract tractor-trailer combination from the Korean Han Jin Company. As contract drivers, they were not allowed to carry weapons. A 2½-ton gun truck led the convoy with the convoy commander, 1LT Dennis C. Mack, riding in the gun jeep which was the third vehicle in the line of march. The second gun truck rode in the middle of the convoy and the last gun truck picked up the rear with the second gun jeep behind it. Besides the gun jeeps, only the last gun truck had a radio.[132]

At 0900 hours, an enemy force estimated at two rifle companies attacked the convoy 300 meters west of a Bridge 27 just as it was passing by Pump Station #3 at the base of Mang Giang Pass. The enemy fired rockets, mortars and automatic weapons into a kill zone that extended the length of the entire convoy. The lead gun truck and convoy commander in the gun jeep raced back into the kill zone while the rear gun truck and gun jeep raced forward to provide cover for the stopped vehicles. Those vehicles that could drove through the kill zone. A mortar round destroyed one tractor-trailer combination and its cargo of construction material, while six other tractor-trailer combinations were damaged and abandoned in the kill zone. Four others were damaged. 1LT Mack exposed himself to enemy fire and moved along the convoy urging the drivers to keep moving out of the kill zone to the top of the Mang Giang Pass where emergency repairs were made and they proceeded on to Pleiku. The rear gun jeep was hit, killing PFC Earl C. Wilson and wounding SGT Nicholas Capizola and one other passenger. The other three wounded were drivers of task vehicles. Mack provided aid and comfort to the wounded and personally supervised their evacuation.[133]

Three APCs between Bridges 27 and 29 returned fire, but it took 10 to 15 minutes for the reaction force to arrive. Two M42 dusters with twin 40mm cannons were the first reaction force on the scene. No gunships arrived during the ambush since they were refueling at An Khe. A platoon of APCs which was behind the 88th Transportation Company convoy came up, swept the area and found five enemy dead. Three of the five American wounded were medevaced to An Khe while the other two, who were slightly wounded, continued on with the convoy. The six vehicles abandoned in the kill zone were later policed up by the commander of the 88th Transportation Company in An Khe.[134] LT Mack received the Silver Star Medal for his actions.

31 August 1968

At 0900 hours on 31 August 1968, a convoy of 33 tractor-trailer combinations under the control of the 88th Transportation Company departed An Khe for Pleiku along QL19. Similar to the last convoy ambush, three gun trucks and two gun jeeps escorted the convoy. At 0945 hours, an estimated enemy force of 100 NVA attacked the convoy with mortars, rockets, automatic weapons, small arms and hand grenades below Mang Giang Pass.[135]

There were two APCs in the vicinity of the ambush site and artillery fire was on target in approximately one minute. It took approximately 10 to 15 minutes for a reaction force of five APCs and six helicopter gunships to arrive. The two gunships normally over the column were on the ground in An Khe being refueled. The force was in contact for approximately 30 minutes. They discovered six enemy soldiers killed.[136]

The convoy had one driver killed and six wounded. Four of the six wounded were medevaced to An Khe while the other two, with minor wounds, continued with the convoy. One gun truck and seven task vehicles became damaged and abandoned in the kill zone. Another task vehicle was badly damaged but able to proceed through the kill zone to a secure area at the top of the Mang Giang Pass. The eight vehicles abandoned in the kill zone were later recovered by the 88th Transportation Company from An Khe.[137]

The engines overheated on ten 5-ton tractors hauling trailers that had cleared the two kill zones while waiting because small arms fire had punctured their radiators. The 88th requested assistance from the 560th Maintenance Company (Direct Support) to develop a device to protect radiators from small arms fire. The mechanics constructed an armored grill that held up against M-16 and M-14 fire.[138]

Pandamonium as it looked around late 1970. It was one of several gun trucks built by the 88th Transportation Company.
Photo courtesy of John Brown and James Lyles

Vietnam War 59

3 January 1969

On 3 January 1969, LT Frank, 88th Transportation Company, 27th Transportation Battalion led a convoy west from Qui Nhon to An Khe in the afternoon. The convoy consisted of nine 5-ton tractors and trailers one low boy, a 5-ton bobtail tractor and seven Han Jin (Korean) cargo trucks which fell in at the end of the convoy. The escort consisted of two gun trucks and the convoy commander's gun jeep in the rear. The convoy was ambushed at 1725 hours approximately one kilometer west of Bridge 19. The enemy had laid an L-shaped ambush with approximately eight soldiers on one side and three on the other firing rockets and automatic weapons. The kill zone stretched approximately 500 meters long. The American drivers drove out of the kill zone with only five 5-ton tractors and one gun truck damaged by fire. On the other hand, the enemy hit three Han Jin trucks killing one driver and seriously wounding the other two. Frank's gun jeep drove up to render aid to the Han Jin trucks and was hit by a B-40 rocket fired from an RPG, killing the driver and wounding LT Frank and his machine gunner. A gun truck drove into the kill zone and laid down suppressive fire. Three of the gunners were slightly wounded and an APC from Bridge 19 came up and fired at the enemy position on the south side of the road. A gun truck and Quad .50 from the lead half of the convoy that cleared the kill zone also drove back and suppressed the automatic weapons and rockets on the south side of the road. MPs sent a gun jeep and a V100 armored car into the kill zone. The gun truck with the three wounded gunners then turned its fire against the small arms fire on the north side of the road. Within 20 minutes, gunships arrived. Artillery was also called in on the enemy positions. The dust-off arrived and evacuated the most seriously wounded. In total, one US driver was killed and five wounded, one Han Jin driver was killed and two wounded.[139]

Lesson

The enemy had varied their tactics in terms of what weapons they initiated with, but they usually initiated the ambush with the most lethal weapon whether it was a mine, rocket or mortar to stop the lead vehicle then followed up with small arms and automatic fire to destroy the rest of the trapped vehicles. The ambushes were primarily linear ambushes with forces dug in on both sides of the road. They primarily attacked on the slopes of the mountain passes where vehicles had to slow down. The ambushes rarely lasted more than 15 minutes since the arrival of the tanks, APCs and helicopter gunships tipped the balance of fire power on the side of the Americans.

The enemy began to reduce the length of their kill zones to 200 to 300 meters in length. This reduction and the 100-meter interval between vehicles reduced the number of vehicles in the kill zone and the gun trucks, while in their infancy, bought time for the reaction forces to arrive. Few gun trucks had radios and the convoys still had maintenance vehicles bring up the rear.

Because the 88th Transportation Company ran convoys by itself, it ran with an average of 30 task vehicles escorted by three gun trucks and two gun jeeps used for command and control by the convoy commander and assistant convoy commander. The smaller convoys provided easier command and control as compared to the 60 to 70 vehicle convoys of the 27th and 54th Transportation Battalions. Two to three gun trucks seemed the right number to defend the convoy in the kill zone. This would become the new policy for all of 8th Transportation Group.

---

[131] 1LT Roland E. Descoteau, Unit History, 88th Transportation Company, 1 January to 31 December 1967, 25 March 1968; and Bellino, " 8th Transportation."
[132] Bellino, "8th Transportation;" and COL Garland A. Ludy, "8th Transportation Group, Sep 1968 - Sep 1969," n.p., n.d.
[133] Bellino, "8th Transportation."
[134] Ibid.
[135] Ibid.
[136] Ibid.
[137] Ibid.
[138] COL Albert J. Brown, Trip Report (8-69). Convoy Security, US Army Combat Deveopments Command, Liaison Detachment, HQ USARV, APO San Francisco, 96375, 20 January 1969.
[139] COL Garland A. Ludy, 8th Transportation Group."

1969 Doctrine Change

COL Garland Ludy had assumed command of the 8th Transportation Group from Joe Bellino in September 1968 and commanded it until September 1969. He further refined the convoy doctrine. Knowing that air support or reaction force would arrive within ten minutes, the enemy would hit fast and hard then withdrawal before the air support would arrive. Kill zones became smaller, from 300 to 400 meters long so the enemy watched for trucks bunching up. Consequently, Ludy would fly over the convoys in his helicopter looking for flagrant violations of convoy interval. He also required that all convoy commanders be officers so if he found any bunching up, he would find the officer at his company the next morning and chew him out in such a way that he would not forget.[140]

Not only did the enemy tend to reduce the size and increase the intensity of the kill zones but they also began attacking the rear half of the convoy. Ludy also came up with a new Standard Operating Procedure (SOP) for reaction to contact. Any trucks caught in the kill zone were to clear it as before. Those trucks ahead of the kill zone were to keep driving to the next security check point. However, all trucks behind the kill zone should stop and not drive through the kill zone in order to avoid creating a bottleneck. In the past the bottleneck effect resulted in more damage because more trucks entered and became disabled in the kill zone.[141]

By this time, 8th Transportation Group had also changed its convoy SOP to restrict the size of convoy serials to no more than 30 task vehicles which resulted in two convoy serials leaving with a 20-minute interval between them. The gun truck ratio of 1:10 required three gun trucks for a serial of 30 vehicles, but the shortage of gun trucks caused some convoy serials to run with only two. By then all gun trucks had radios so they could communicate with each other and the convoy commanders. This also allowed them to operate more independent of the convoy commander.[142]

The procedure for reaction to contact required gun trucks to enter the kill zone and turn the fight back on the enemy as fast as they could. If any task vehicles were disabled, then the gun trucks would enter the kill zone and protect the vehicles and rescue any wounded or stranded drivers. The gun trucks remained in the companies and were more often assigned to platoons so that the élan of the crews did not cause them to lose their identification with the men they protected. The gun truck crews knew at the end of the day, they had to return barracks and face the men they escorted.

SP5 Steve Calibro, 523rd Transportation Company, 1968-1969, had trouble with new lieutenants as convoy commanders. They came in with little to no experience and took charge of soldiers who had been there for six or eight months and had survived numerous ambushes. These convoy commanders wanted to position a gun truck. Calibro reflected, "When shit is flying, you ain't worried about positioning the son of a bitch." Many gun truck crews felt that because each ambush was different, fights were not won by tactics, but superior fire power. "You just went in there and open fire on everything you can see or cannot see. You want to throw as much lead as you can. Half the time we did not see them. We wanted to unleash with everything we had and as much as we could put out there at one time." They used spraying fire to suppress the enemy because half the time they could not see the enemy, which fired from the tree line, from behind rocks or even behind a water buffalo. Once after an ambush had ended, Steve looked over and saw a water buffalo with six legs and realized a Viet Cong soldier was hunched over walking behind it. The convoy commander was telling them to move on. Calibro said, "If you move on, somebody is going to get shot." The gun truck crews felt the lieutenants did not always know what was going on and their tactics did not match the situation.[143]

All the gun truck crews interviewed felt the same way as Calibro about tactics. Officers wanting to discuss tactics offended most gun truck crew members. In the Second Golf War, 2003-2011, however, major emphasis was later placed on the development of tactics and the placement of gun trucks in the convoy and kill zone by the officers. LTC Edward McGinley, commander of the 24th Transportation during a tour in Kuwait, 2005-2006, explained that the gun truck crews in Vietnam had to face every kind of threat, whether mines, small ambushes or large complex ambushes. Having to face every kind of threat, they could not study or develop tactics for everything, but the convoys driving in Iraq after 2005 faced primarily one type of attack – mines then better known as improvised explosive devices (IED).

With one primary threat to contend with, the leaders could analyze patterns and develop tactics accordingly. This was not the case during the Vietnam War though.

When COL Ludy assumed command of the 8th Group, most gun trucks were 2½-ton trucks, some armed with a ring-mounted M-2 .50 caliber machine gun on the cab and two handheld or pedestal-mounted M-60s in the gun box. There were a few 5-ton gun trucks with APC hulls added to the bed of the trucks. By then all gun trucks also had names and seemed to take on personalities as the design of each gun truck was left to the original crews. Each gun truck had a different configuration of armor. The gun truck's name reflected the character of the crew and the truck. The crews were all volunteers selected by the former crews from among the best drivers. The crews began to improve their gun trucks like an infantryman improved his fighting position. They switched from 2½-ton trucks to 5-tons which would allow for more armor, and they began to replace the M-60s with heavier machine guns. Ludy also had armor plating added to the doors of task vehicles.

Ludy, however, did not allow any elaborate decoration of the gun trucks. All had to be painted olive drab so they would blend in with the other task vehicles but he allowed the crews to stencil the names of their gun trucks on the side of the gun box, but some went a little further. Some cheated and added elaborate art work but it was rare. After COL Ludy gave up command of the 8th Group in Serptember 1969, the crews painted elaborate art work on the sides of their gun boxes and added varnish to the olive drab paint to make the color darker until they just painted the gun trucks black. From then on, ambushes would be described in terms of the gun trucks as if they were living things.

523rd Light Truck Company, Eve of Destruction and the élan of the Vietnam Gun Truck

The names reflected the spirit of the crews and as the art work became fancier, it reflected their élan. The crews accepted greater risk in that when other drivers wanted to escape the kill zone, the gun trucks had to drive into the kill zone. Fortunately, every company had men who measured up to that challenge. Many of them had just finished light vehicle operator course and joined their company in Vietnam with less than a year in the Army. At that time promotions came fast and in a year, a 19-year old kid could be an NCO in charge (NCOIC) of a gun truck making life and death decisions for his crew and the rest of the convoy.

SP4 Steve Calibro arrived in the 523rd Transportation Company in June 1968 and picked up a 2½-ton gun truck with no name. By then the 523rd had already encountered ten major convoy ambushes that year. Deuce-and-a-halfs loaded with armor plates and sandbags, however, were very slow going up the mountain passes and could not keep up with the prime movers. As the new gun truck NCOIC, Steve wanted a more powerful 5-ton gun truck. The company had 5-ton cargo trucks and the few 2½-tons were administrative vehicles. So Steve supervised the building of the new gun truck. He had the original gun box lifted off the deuce-and-a-half and put on the bed of a 5-ton. They then stacked up a single row of sandbags about two feet or knee high along the inside of the box with perforated steel planking (PSP) welded at the corners into an interior wall to hold the sandbags in place. Essentially the gun truck had a single wall of ¼-inch steel plate that would only stop AK rounds if they hit at an angle. The crew then placed the M-134 mini-gun ammunition cans along the walls on the sandbags and then surrounded them with sandbags. The cans held 4,000 rounds of 7.62mm for the M-60, which at that time were still handheld since they did not have any pedestal mounts for them.[144]

Barry McGuire's antiwar song, "Eve of Destruction," was very popular at that time and Calibro named his new gun truck after it. A guy in the motor pool painted the name on the OD gun box in Old English style white letters with the red shadow. The same guy also painted Corps Revenge. Although COL Garland Ludy, who took command of 8th Group in September, did not like elaborate art work the Eve had no complaints from higher. They put the Eve of Destruction on the road 4 July 1968. At that time QL19 was still a dirt road all the way up An Khe Pass and Pleiku so trucks were still hitting mines. Gun trucks would have to pull over and secure the

Corps Revenge parked next to Eve of Destruction.
Photo courtesy of James Lyles

True Grit was later renamed Black Widow.
Photo courtesy of the US Army Transportation Museum

King Kong. Photo courtesy of the US Army Transportation Museum

damaged vehicle until a wrecker came up to recover it.

By that time, Calibro remembered the 523rd only had three gun trucks; Grim Reaper, Eve of Destruction and King Kong. King Kong was an APC gun truck. The APC gun truck was a death trap. There was no access from the cab to the APC hull as with a gun box so if the driver was wounded, no one could get to him. He would be helpless and no one could take over the driving. Initially the APC had only one .50 on the turret and the other guy in the hull was an ammo bearer.[145] SGT John Burnell later rebuilt the Grim Reaper in December 1968 after it was destroyed in an ambush. He renamed it Corps Revenge after Ray "Corp's" Hastings who was severely wounded on the gun truck, 54th Best (formerly the Gamblers), in the 13 November 1968 ambush. A single layer of steel was bolted to the outside of an M-54A1 5-ton cargo truck with sandbags built up to the height of the bed of the truck.[146]

After a couple weeks, Calibro found a tripod on a jeep in the motor pool and decided to put a .50 cal on it in the middle rear of the gun box so he could fire rearwards, forward and to the sides. He set the tripod up and somehow found two .50s (that he was not supposed to have). Steve then took them down to the motor pool and had a guy weld a mount that would hold two .50s. Jerry L. Bailey was the driver and the gun truck still only had two gunners, newly promoted SP5 Calibro on the twin .50s and another M-60 gunner.

By the end of 1968, the 523rd had a total of five gun trucks, Eve of Destruction, King Kong, Corp's Revenge, Ace of Spades and The Filthy Four. The 5-ton gun truck, Ace of Spades, was an identical twin of Eve with the same gun box design except it did not carry spare tires. Each had a driver, two gunners and a grenadier.[147]

COL Ludy felt the placement of the gun truck in the lead made it more vulnerable to land mines and the enemy usually initiated the ambush on the gun tuck. So in November 1968, he replaced the lead gun truck with a reconnaissance jeep, which would scout the terrain ahead of the convoy for signs of enemy and obstacles. A task vehicle then led the convoy serial and the gun truck varied its position in the march serial daily so the enemy could not anticipate its location in the convoy.[148]

A staff sergeant had been bugging SP5 Calibro to go out on the Eve for a ride. Steve did not remember what his job was, just that he always walked around the company area. On 13 November 1968, Calibro

After COL Ludy left in September 1969, the art work on the gun trucks became more elaborate.

finally told him he could go as a "ride-along" on the run to An Khe. Steve warned him, "You're an E6 and I'm an E5; before you get on you need to understand who is in charge of this truck. I'm taking no shit from you. It's my truck. I'm in charge." He acknowledged, "Okay, fine."[149]

The convoy made it to An Khe without any problems. They went up with about 40 trucks but picked up more that had been waiting there over night. They may have had trouble unloading and were stuck overnight. The convoys to An Khe often returned with more trucks than they started. Calibro's convoy was returning empty with about a hundred trucks and was almost at the bottom of the Pass, when the staff sergeant tapped the driver on the top of the head and told him, "You must be getting tired. Let me drive for a little bit." They switched and the staff sergeant was not in the driver's seat for five minutes when a rocket hit from the left-hand side right between the box and the cab. The explosion tore the staff sergeant's legs up with shrapnel, blew the armor off the door, scattered the driver's seat, blew the glass out and blew the 2X6-inch wooden plank that went around the top and held the armor plating together back into the crew in the box. The truck drove off the road down into the grass on the right-hand side. Calibro yelled at the driver to get the truck out of there, but the staff sergeant was seriously wounded. The truck came to a stop in the grass about ten feet from a culvert that ran underneath the road near the Korean fire base. As Calibro stood up and turned around, he saw about 60 to 70 NVA pouring out of the culvert. They approached to within a couple of feet of the Eve and then the crew started firing as fast as they could. It looked as if the NVA were going to come over the top of the truck when Steve heard gun fire. He looked up and saw a Quad .50 (possibly Bounty Hunter) come up and NVA scattered. The fight lasted so long that the whole bottom of truck was ankle deep in brass. He had even melted the barrel of the .50. and at that time he had no idea what it looked like when a barrel melted. He saw the red tracers floated up and down like a feather instead of flying straight. He thought that was weird, and wondered what was causing it to happen. He also double loaded an M-79. He fired a round and then heard "bloop," loaded another round and heard "bloop," then shoved third round in and it would not fit. He had two rounds stuck in the barrel.[150]

Steve did not see or remember what other gun

Ace of Spades had the same rear gun mount configuration as Eve of Destruction. Photos courtesy of the US Army Transportation Museum

trucks came into the kill zone because of the intensity of the fight and he was calling for "Dust off" on the radio at the same time. Calibro and his driver were wounded and the staff sergeant was killed. Only one gunner on the Eve was not hit. A helicopter medevaced the wounded out.[151]

The company motor sergeant worked all night to rebuild the Eve. He lifted box off with a crane and put it on another 5-ton, put the armor on the doors and painted the same bumper number as the original truck. He wanted to put it back out on the road as if nothing had ever happened to it. Calibro was off duty for a week before he returned to his gun truck.[152]

22 January 1969[153]

On 22 January 1969, the Eve of Destruction was escorting an empty convoy back. They had just cleared the bottom of the Pass and were about a mile down the road. The Eve was the tail gun truck and had a radio. SP5 Calibro heard another convoy behind his get hit at bottom of An Khe Pass. The Eve was the only gun truck from the lead convoy to go back to the kill zone. He saw four cargo trucks burning and four or five drivers laying on the road bleeding. There were also three to four gun trucks from that convoy. The ambush lasted a long time and Calibro had fired so many rounds through his twin .50s the barrels were hot. After the fight he pointed the barrels skyward and slipped in the pin to hold them up and the weapons kept firing. An officer (possibly a captain) who was the convoy commander yelled up at him, "I said cease fire!" Steve could not do anything about it because the rounds were cooking off. The officer wanted him to open the feed tray and pull out the belt. That could have caused a round to explode in his face so he told the officer, "Fuck you, if you want to open it up come up here and open it up, but I'm not going to." Meanwhile, fighter jets then flew in. For entering the kill zone, SP5 Calibro and his crew, SP4 James E. Pulley and PFC Jackie W. Witten, earned ARCOMs with V device.[154]

14 February 1969

A convoy of thirty-four 5-ton cargo trucks, seven 2½-ton cargo trucks, five Han Jin (Korean contract) reefers, three Han Jin dry goods vans led by 1LT Jay M. Shelley, 512th Transportation Company, headed west from Qui Nhon toward Pleiku. The convoy was escorted by four gun jeeps and five gun trucks of the 523rd spaced evenly throughout the convoy. Eve of Destruction was the first gun truck and at 0925 hours, SP5 Calibro saw

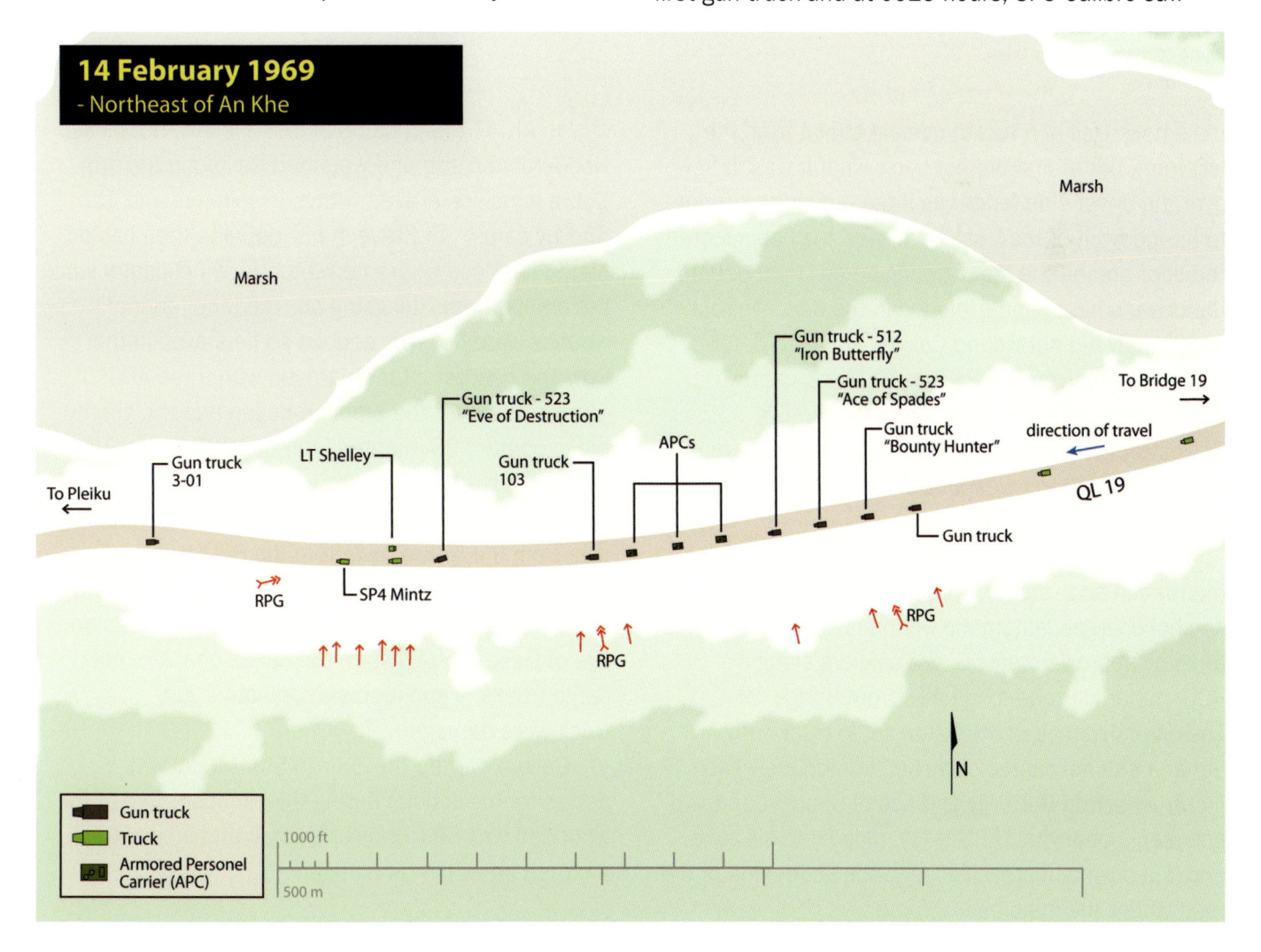

14 February 1969
- Northeast of An Khe

Vietnam War

Iron Butterfly with two rear side mounted .50s and two forward side mounted M-60 machine guns.
Photo courtesy of James Lyles

the smoke trail of a rocket coming at him from the left front, but at first did not know what it was. It flew over the driver side fender as if the enemy was trying to hit the front of the truck. It missed the cab, cleared the door and hit the steel plating on box right behind the driver's head, and exploded in the box. The rocket surprisingly did not wound Calibro. The lead 5-ton cargo truck, driven by SP4 Gary C. Mintz, came to a stop, causing the next 5-ton and the Eve of Destruction behind it to also stop. The gunners in the Eve swung their weapons around and engaged the enemy.[155]

The convoy was attacked from both sides of the road in an L-shaped ambush with rockets, satchel charges, automatic weapons and small arms by approximately 50 NVA soldiers along a 400-meter kill zone a half mile from Bridge 19. The enemy damaged the 5-ton cargo truck with either a satchel charge or rocket. Immediately two gun trucks, Iron Butterfly and Ace of Spades, and two gun jeeps, one with 1LT Shelley, raced into kill zone and laid down suppressive fire to the south side of the road where the main enemy force was.[156]

An AK-47 round came across the left .50, nicked wooden hand grip and wounded Calibro in the arm. Cobra gunships arrived within 10 minutes and 1LT Shelley came, put Steve in his jeep and then hauled him to fire base where he waited for a helicopter while the ambush was still going on. The other gunner's wounds were not very serious so he stayed in the Eve. The nearest infantry arrived with three APCs 15 minutes after the ambush began and the enemy broke contact 25 minutes later at 1005 hours. The ambush lasted 40 minutes with one US driver, SP4 David Leroy Bell, killed from the 512th, eight wounded (five from the 523rd, two from the 669th and one from the 512th) and five confirmed enemy killed. In addition to the initial damage to the 5-ton cargo and Eve of Destruction, two 5-ton cargo, one 2½-ton cargo trucks, a gun jeep and one other gun truck were damaged.[157]

Calibro earned the Bronze Star Medal with V device for his actions during this ambush. One of the wounded was SP4 Bobby Newman and in less than a month he would be hit again.[158]

4 March 1969

LT Hutchins, of the 512th Transportation Company, led a 54th Transportation Battalion convoy to Pleiku and back. The convoy consisted of twenty-seven 5-ton cargo trucks, ten 2½-ton cargo trucks, three 5-ton tractors and trailers, and one maintenance truck. Because this serial contained 41 task vehicles, it was escorted by four gun trucks and one gun jeep. On the return trip, Hutchin's convoy march serial followed behind a convoy of 28 cargo trucks led by LT Shelley of the 523rd.[159]

Hutchins' convoy had just come down An Khe Pass and then turned north just before the turn onto the coastal plain. The road leveled out with the mountain on the west side of the road and a river following it on the east side. At 1730 hours, the enemy was concealed on both sides of the road and initiated the ambush with a B-40 rocket fired at the lead cargo truck, which missed. This truck and the next four cargo trucks received small arms fire while they cleared the kill zone. The sixth vehicle, a gun truck, stopped in the kill zone to return fire and two additional gun trucks drove forward to assist in the kill zone while the remainder of the convoy did not enter the kill zone. Helicopter gunships arrived overhead within minutes and gun trucks from lead convoy serial also joined the fight. A total of five vehicles were damaged and ten soldiers wounded. A sweep of the kill zone produced three enemy killed in action.[160]

The new convoy organization reduced the size of the convoy serial but gave each serial at least two gun trucks and a gun jeep for protection. So once the call for contact went out, gun trucks from other serials would race into the kill zone to assist. Eight gun trucks eventually converged in the kill zone looking for a piece of the action. While the convoys may have been strict about the interval while traveling, the gun trucks did not worry about it when they converged in the kill zone. Eight gun trucks occupied a space of 300 meters less than 40 meters between gun trucks - a lot of fire power in a small space.

The Eve of Destruction was rocketed four different times and had to be rebuilt each time. Steve never thought about it at the time but remembered years later, the convoy would pull over for a break, and the Vietnamese would walk up and down selling sodas. A few times they took pictures of the Eve and then the next day it would be hit about the same place they had stopped. Veterans told Steve that the Eve was almost always in the middle of the ambushes and never pulled away.

SGT John D. Burnell, NCOIC of Corp's Revenge, remembered his convoy passed the old petroleum pumping station at the bottom of An Khe Pass,

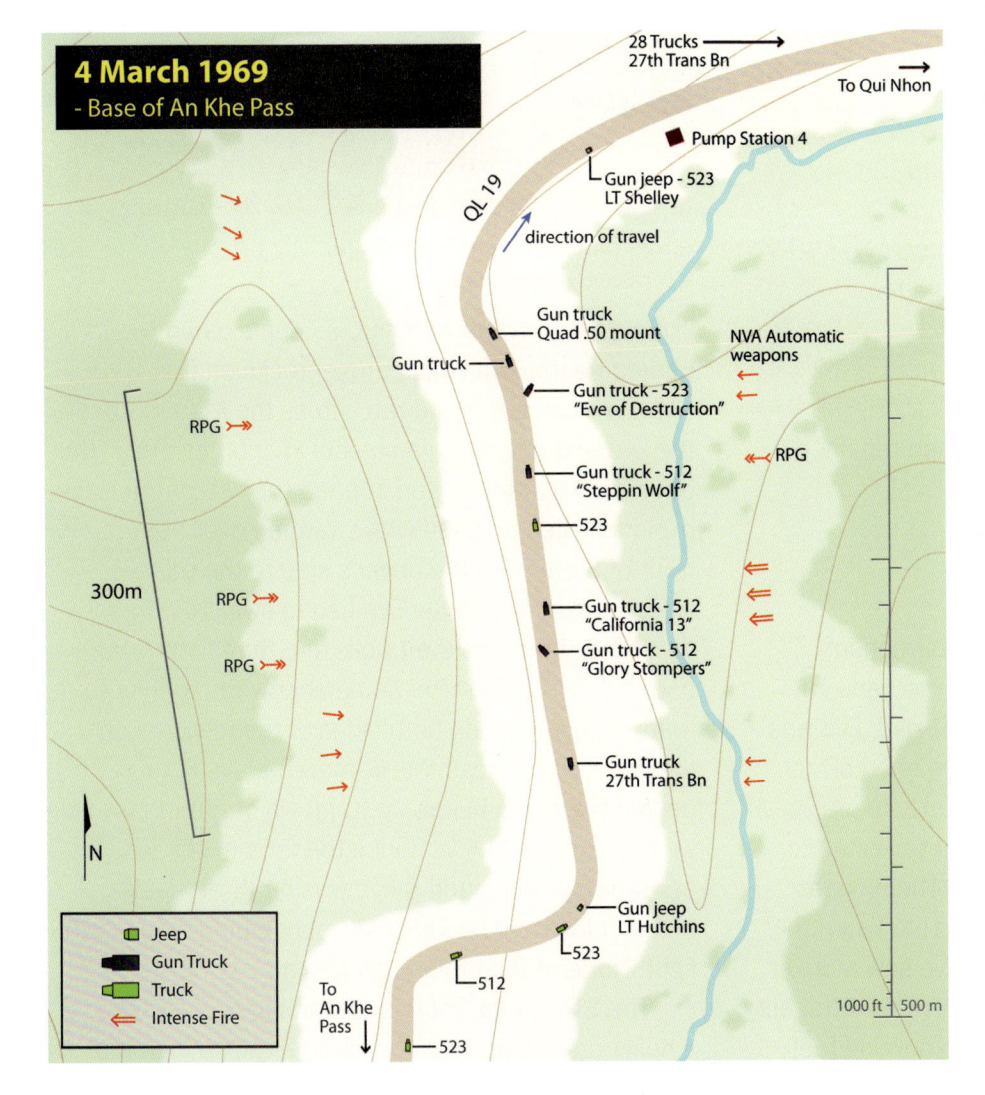

Vietnam War 67

which was guarded by four or five M42 Dusters of the 4th Battalion, 60th Artillery. The Eve of Destruction was the first gun truck and Corp's Revenge was second in the middle. Burnell noticed something about 200 yards to his left and looked. A group of VC charged out of the tree line straight at the Dusters so fast the tracked vehicles did not have time to turn their twin 40mm anti-aircraft cannons on them. Burnell also saw the Eve halfway up the pass disappear over the crest of the hill. Burnell popped a red flare and then told his driver to turn into the attack. The gunners fired at the VC while some of the Duster crews ran toward the gun truck. The fight was getting desperate when all of a sudden, Burnell saw the Eve heading toward the fight and he knew everything was going to be all right. The combined firepower of the two gun trucks drove the enemy back into the trees. The crew of the Eve had been watching for Corp's Revenge to crest the hill, but when they saw the flare and heard the gun fire, they raced to the sound of the guns. The Eve had a reputation of being in all the fights.[161]

Corp's Revenge was destroyed by an RPG in an ambush that killed its driver, Robert C. "Herm" Herman on 18 April 1969. The crew then renamed the gun truck, Herm's Revenge.[162]

24 May 1969

On 24 May 1969, the Eve of Destruction escorted a convoy with some fuel tankers near "VC Village" almost into Pleiku. Eve was the first truck hit with a rocket and it coasted off to the side of road. The NVA then hit about four or five cargo trucks in front of Eve and the convoy came to a dead stop in the middle of the road. The drivers got out and returned fire with their M-16s. When the Eve rolled off the side of the road, Steve saw a driver lying in the ditch firing his M-16. The NVA fired tracers into fuel pipeline and caused it to leak burning fuel

Corp's Revenge was renamed Herm's Revenge.
Photos courtesy of John Burnell and James Lyles

like a flame thrower right into a couple cargo trucks parked nearby, which set those trucks on fire. The NVA then came out and ran across the road in front of the Eve and one threw a satchel in another truck. The driver had already gotten out and the cab of the truck exploded. Because the NVA swarmed over the road in front of them from the driver's side to the passenger side, the gunners of the Eve had to swing their guns around and fire into the trucks and then fired off of both sides of truck after some enemy ran across the road. With enemy on both sides of the road they had the convoy in a cross fire. Some were even hiding behind the cargo trucks up front. No other gun truck was hit except the Eve. Firing the .50s exposed him from the waist up, and Steve was hit three times in the shoulder and chest by AK rounds but stayed in the fight. None of the other crew was hit.

The 8th Group Commander would fly overhead in a helicopter inspecting convoy interval, so high that the drivers could only hear the rotor wash but not see the bird. During the ambush they had wounded and burning vehicles and the colonel was talking to them like he was on the ground. This pissed Calibro off and he thought, "I could shoot this fucker down with these .50s if he would just get close enough." Five minutes after the kill zone was secured the colonel landed. After the ambush Calibro was medevaced to Pleiku earning his third Purple Heart Medal. His driver, PFC Thomas G. "Tom" Johnson, later told Steve he thought he was dead. The driver had not seen so much blood in the truck. 1LT Ronald C. Loveall and SP4 Vernon L. Branham received Bronze Star Medals with V Device for that ambush.[163]

On 28 May 1969, Steve Calibro was sent home due to wounds. He had 11 days left in country. He was flown into Japan, then Travis, and finally Letterman Army Hospital in San Francisco because he was from Concord, California.

In June 1969, CPT Joseph McCarthy II assumed command of the 523rd Light Truck Company in a state of low morale. He only had about a platoon's worth of drivers available for the road since the rest had less than 30 days to rotate home. He had to relieve his first sergeant within the first two weeks and picked up 80 new soldiers which he assigned to 2nd and 3rd Platoons. He dumped all his bad apples in the 1st Platoon.[164]

By then, SP5 Roger D. Champ had arrived in Vietnam but was initially assigned as COL Garland Ludy's clerk typist but did not like the job, so after a couple weeks, he was transferred to the 523rd Transportation Company. He impressed CPT McCarthy as an ass-kicker and was assigned as platoon sergeant of 1st Platoon and promoted to acting sergeant. McCarthy assigned SGT Larry Cole, who arrived in September, as 2nd Platoon Sergeant and SSG Lee as the 3rd Platoon Sergeant, and tasked both to built gun trucks. The Eve of Destruction was about the only functioning gun truck left in the 523rd and had to cannibalize other trucks to stay on the road. Champ became the NCOIC of the Eve. McCarthy named Cole's gun truck, True Grit, after a John Wayne movie, and Lee's gun truck, Uncle Meat, named after a rock album by Frank Zappa of the underground blues band, "Mothers of Invention." It was a kind of a head

CPT Joseph McCarthy, was a fan of Frank Zappa's underground blues band and named this gun truck after Zappa's Album.

Eve of Destruction.
Photos courtesy of the US Army Transportation Museum

Black Widow with early art work. Photo courtesy of James Lyles

Vietnam War 69

group and McCarthy did like his parties, but none of the upper brass ever made the connection. Uncle Meat was a brand new gun truck and True Grit was renamed Black Widow shortly after McCarthy left January 1970.[165]

4 January 1970

A westbound convoy left the marshalling yard at Cha Rang Valley on the morning of 4 January 1970 bound for An Khe and Pleiku along QL19. The convoy was divided into two serials of no more than 30 vehicles. Most ambushes took place in either of the two passes, An Khe or Mang Giang. Drivers felt rather safe in the coastal plain since the Republic of Korea (ROK) Tiger Division secured the area.[166]

Three gun trucks escorted the lead serial of a 27th Transportation Battalion westbound convoy. The convoy commander, LT James Himburg, rode in the front of the convoy jeep had twin 7.62 mm M73s machine gun. SGT Terrance Van Brocklin was the NCOIC of the 512th Transportation Company 5-ton gun truck, Lady Be Good (former Sergeant Pepper II), which was lead gun truck. The middle gun truck came from the 27th Battalion with a crew thrown together for that mission. SGT Terrance Van Brocklin knew the NCOIC because he had ridden with the 27th before. Cold Sweat, from 669th Transportation Company, was the last gun truck with Parson as the NCOIC and the gunner was Pennington.[167]

The second serial of the 54th Battalion escorted by two gun trucks and one gun jeep followed 20 minutes behind the first. SGT Champ was the NCOIC of the Eve of Destruction in the following serial. The Eve was a double-walled 5-ton gun truck with two forward left and right .50s and dual-mounted .50s on a single pedestal in the rear. SSG Lee, 3rd Platoon Sergeant, was the NCOIC of Uncle Meat, which brought up the rear. Uncle Meat was also a standard 5-ton gun truck but with three .50s, two forward and one rear.[168] The weather was clear which would not impede aerial support or medevac.

The lead serial reached Hogson village south of Bridge 10 just below An Khe Pass. A couple hundred yards away on the south (left) side of the road was a small ridge surmounted with a Buddhist temple. The ridge dropped off into a rice paddy that stretched north of the road.[169] Because this area was secured by the Koreans they never expected to be attacked. The enemy let the first part of the convoy pass and then someone in the village fired an RPG hitting a

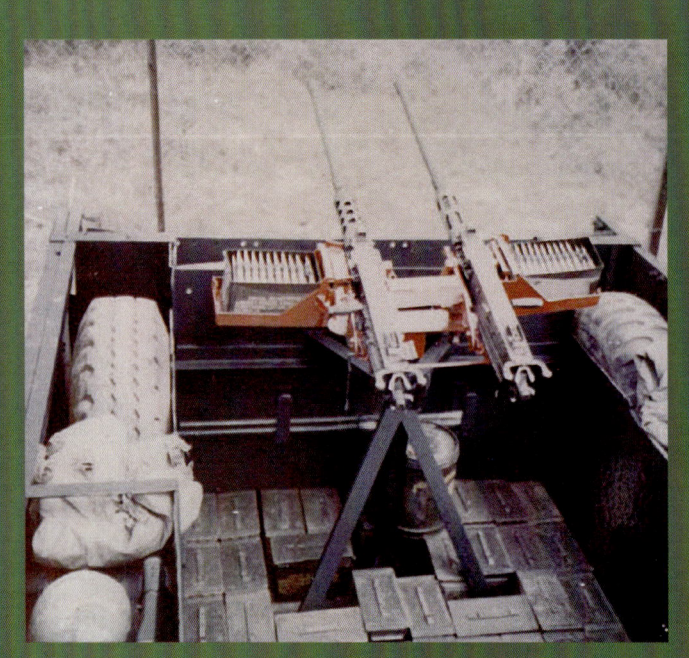

Rear view of gunbox of Eve of Destruction showing double wall and dual .50s. By this time, gun trucks carried as many ammo cans as would fill the bed of the truck.

Forward view of the gun box of Eve of Destruction showing radio and side mounted .50s. Cold drinks were kept in the marmite can.
Photos courtesy of the US Army Transportation Museum

70   Convoy Ambush Case Studies - Volume I

glancing blow off the driver's door of the middle gun truck;

but it blew the hood half way off hanging on only by one of the hinges on windshield. Shrapnel also peppered the fender. It was followed by small arms fire. Van Brocklin did not remember the middle gun truck calling "Contact," but the driver kept driving out of the kill zone.[170]

Cold Sweat then moved up from the rear and probably called "Contact." Himburg was ahead of Lady Be Good when Van Brocklin heard the firing and told his driver to turn around go back into kill zone. Lady Be Good then met the gun truck from the 27th and blocked the road to stop them. The driver of the second gun truck had been a cargo driver for the last 11 months and never been ambushed. He told the crew of Lady Be Good the VC were all over the place. Lady Be Good then drove ahead into the kill zone.[171]

As Lady Be Good came around corner, Van Brocklin could see Cold Sweat firing and there were bodies on the road. Lady Be Good then fired at an extreme angle. They fired at everything not seeing any enemy or where the firing came from. The enemy fired no more rockets but lots of small arms fire. Lady Be Good then focused its fire at a brick building in the village. Van Brocklin could then recognize the bodies on the road were Vietnamese civilians not dressed like Viet Cong. There were only two gun trucks in kill zone at that time and they fired off more than a couple boxes of ammo before they lifted (ceased) their fire to listen. At that point LT Himburg's jeep drove down into the kill zone. It seemed to be quiet for a moment and then Van Brocklin saw a man in black PJs run to the right from the edge of the village across an opening to the hill behind it. That was the only VC Van Brocklin identified and hit him while he was still in the grass.[172]

As LT Himburg continued into the kill zone small arms fire opened up on his gun jeep. The enemy fire knocked out his radio and machine guns, but miraculously did not hit Himburg or his driver, Rippee.[173]

At that moment the gun truck from the 27th returned. Everyone continued firing, which ended the fight pretty quickly. SSG Ronald Britton, the NCOIC of convoy, escorted the lead part of the convoy that was ahead of the kill zone on up the Pass. The gun truck crews discovered the wounded civilians on the road were women and LT Himburg gave Van Brocklin

Lady Be Good. Photo courtesy of James Lyles

Cold Sweat - In 1969, COL Ludy had the yellow strip painted on the nose as recognition of 8th Group trucks. The 523rd added the extra stripe after it went north in 1971.
Photos courtesy of the US Army Transportation Museum and James Lyles

This 20 November 1970 ambush occurred near Bridge 5 on the Coastal Plain, but the terrain closely resembled the ambush Bridge 10 on 4 January 1970. Photos courtesy of Larry Wolke

the frequency to call in a dustoff for the civilians. By policy, the American helicopters did not medevac Vietnamese civilians, but Himburg ordered it anyway. Van Brocklin respected the lieutenant for that decision. Van Brocklin called in the dustoff, and they asked where they were, were there any NVA, and asked if they had smoke to mark the landing zone. Van Brocklin threw a smoke grenade and the medevac helicopter landed.[174]

Van Brocklin had previously acquired three stretchers to sleep on if they had to rest over night (RON) at their destination. LT Himburg and Van

72   Convoy Ambush Case Studies - Volume I

Brocklin carried both women to the chopper on the stretchers. As the helicopter lifted off, they were immediately mortared from the hill. Once they had everyone accounted for, Himburg climbed on Lady Be Good to call in the preliminary report. The gun trucks then drove up the Pass to An Khe. At that point they switched frequencies since another unit had responsibility for securing that part of the road. The lead serial had no wounded.[175]

At that time they had no contact with the convoy serial behind them. Van Brocklin's convoy had not called for help and did not know they were coming, so they knew nothing of what happened when that convoy entered the kill zone. They did not remember seeing the other two gun trucks when they arrived and considered the ambush over.[176]

Upon hearing "Contact," the Eve of Destruction and Uncle Meat raced full throttle toward the kill zone. By the time they arrived the fight had gone on for half an hour or more. SGT Roger Champ remembered seeing several task vehicles, both gun trucks and the gun jeep were shot up. The two new gun trucks maneuvered around the disabled vehicles. Evidently the ambush started over again because mortar and rocket fire fell like rain drops but fell short in the rice paddy. This was the worst ambush that SGT Champ had ever seen. As soon as he reached the kill zone, he called in helicopter gunships from An Khe, but they reported that they were "socked in" by weather. Champ saw no clouds in the sky.[177]

The Eve nosed itself into position at a 45 degree angle in the kill zone and stopped behind a disabled vehicle. This allowed all machine guns to fire forward. SGT Champ saw puffs of smoke in the temple and fired at it. A 2½-ton truck from the 4th Infantry Division (Mechanized), which had infiltrated the convoy, tried to run the kill zone. As it maneuvered around the Eve off the road on the north side, a B-40 rocket disabled it wounding the driver. Meanwhle the gunners fired continuously.[178]

They had been in the kill zone for about an hour when a mortar round hit near the edge of Champ's gun truck and shrapnel blew a gunner, Tony Inuchi, out of the gun box onto the pavement between the gun truck and disabled truck in front of them. Champ jumped down to rescue him and examined the gunner's wounds. The gunner had worn his flak vest open and received a sucking chest wound. As Champ bent over to administer first aid, pieces of asphalt began hitting him in his face. For the first time he realized that they were also receiving small arms fire. When he looked up he assumed that the only place the small arms fire could come from was the banana grove that ran along the irrigation ditch a couple hundred yards away. Champ then dragged the wounded gunner around to safety behind the truck. He also dragged a wounded Vietnamese man who unfortunately was caught in the kill zone. He then climbed back into the gun box and directed the other gunners to fire into the banana grove. Suddenly enemy soldiers rose up and began running away. To their shock, the Koreans had secretly positioned themselves behind the enemy and cut them down. This broke the enemy attack. According to his award citation the ambush had lasted three hours.[179]

A Huey gun ship finally arrived at 1000 hours after the fighting died down. Champ had put in a call for a medevac but none would respond. He then called down the Huey gunship and loaded the wounded gunner, a Vietnamese man, and the wounded 4th Infantry Division soldier in the Huey. The Infantry driver told Champ that he had just reenlisted. Since all the tires of the Eve were shot out, a tire truck came up and the crew of the Eve replaced the tires. Any trucks that could not be repaired on the road were hooked up to a bobtail (tractor without a trailer) and towed.[180]

When the lead serial arrived at Pleiku, LT Himburg had Parsons and Van Brocklin accompany him to fill out supplements to his official ambush report. They reported to an underground bunker. Someone must have closed the road that day so his convoy returned to Qui Nhon the next day.[181] When the convoy rode past the same area, the Koreans had stacked all the enemy dead alongside the road to show the truck drivers, a common practice of the Koreans. They were proud and liked to show the Americans their kills before they burned them. There were so many dead that it took a truck load of fuel to later burn them.[182]

CPT Joseph McCarthy II, Commander of the 523rd Transportation Company, submitted SSG Lee and SGT Champ for the Silver Star Medals mainly for picking up Vietnamese rice farmers and loading them into the gun trucks to prevent them from getting killed by VC crossfire during the ambush. Very rarely did truck drivers receive valor awards, so CPT McCarthy wrote a six-inch or thicker award packet to compensate for this drawback and both received the valor medals.[183]

Lesson

The enemy had taken advantage of attacking in an area where the drivers relaxed their guard. The availability of air support and response time of the local combat arms usually limited most ambushes to no more than 15 minutes. The delayed response by attack helicopters and apparent absence of the Korean Tiger Division inspired the enemy to fight longer.

Since the enemy reduced the size of the kill zones they tended to let the lead trucks pass and attack the middle or rear of the convoy. The organization of the convoy into serials of 20 to 30 vehicles kept at least two gun trucks within supporting distance. In most of the ambushes the gunners rarely could identify from where the enemy small arms fire came, so they employed spraying fire in every direction hoping to hit the enemy. Fortunately, SGT Champ was able to make positive identification of the enemy location and directed the fire of the other gunners onto that location. It was a question of superior and accurate fire of .50 caliber machine guns that turned back the enemy and the Korean soldiers that were maneuvering to flank the enemy were in position to prevent the enemy from escaping.[184]

New NCOIC

In February 1969, Harrison "Bud" Whitehead arrived in Qui Nhon and started out as a stevedore in the 205th Transportation Company, but the call went out for volunteers for gun trucks so he joined the 523rd about the time Champ left and became the NCOIC of the Eve of Destruction. The chassis of the Eve had been destroyed by a RPG so they put the gun box on a brand new 5-ton chassis. Lulu's song, "To Sir with love," had been a big hit in 1967, and Bud liked it, so he had "To Charles (revering to the VC) with Love" painted on the bumper as a joke and fastened a rubber hand raising the middle finger on the bumper. By that time, the Eve gun box had two front left and right .50 caliber machine guns as well as the twin .50s mounted on the back. While other gun trucks had three gunners, the Eve still only had two. The front gunner switched between the right and left gun. It mounted an AN/VRC-12[185] radio for communication. While on the Eve, Bud participated in at most three ambushes.

About this time the rumor spread that the enemy had a bounty on gun trucks, probably because they targeted gun trucks first. The kill zones were smaller and more lethal, so only three vehicles might enter a 300-meter kill zone. Therefore it made sense for the enemy to disable a gun truck first, since the next gun truck was probably ten vehicles away. Because some gun trucks like the Eve were hit more often than others, the crews believed some had higher bounties. In response, the crews wanted their trucks to stand out from the rest. Consequently, the artwork became bolder and nearly every truck driver wanted to take a picture of them, which added to the élan of the crews.

In November 1969, Bud Whitehead recruited Gerald Merdutt as a gunner on the Eve. Bud wanted some seasoned veterans on his gun truck, and Gerald had been in the 523rd since September and was slightly wounded in an ambush. After Bud left in February 1970, Merdutt became the NCOIC of the Eve and left Vietnam in September 1970. The next crew of the Eve of Destruction took it north to I Corps in January 1971 for Operation Lam Son 719.

While every other driver's job was to escape the kill zone, the job of the gun truck crews was to race into it. This required a special type of man willing to accept risk to save fellow drivers. They received no extra pay or benefits, only the pride of serving on a gun truck. Consequently they expressed that pride in the name and artwork on the side of their gun truck. Believing the enemy had a bounty on their gun trucks, they had a sense of bravado and wanted their truck to stand out in defiance of the enemy. The gun truck reflected the crew, so the NCOIC ran the gun truck and picked his crew. If anyone did not measure up or hid from a fight, he was removed from the gun truck. So only the best drivers aspired to and manned gun trucks. They had an élan that would stay with them for the rest of their lives.

The quote "To Charles with Love" was a twist on the popular 1967 movie titled, "To Sir with Love." Charles in this case referred to the Viet Cong, commonly referred to us as "VC" or "Victor Charlie" or just "Charlie." Just above the quote was placed a rubber hand raising the middle finger. Photo courtesy of the US Army Transportion Museum

---

[140] Joseph McCarthy email to Richard Killblane, June 27, 2007.
[141] Ibid.
[142] Ludy, "8th Transportation Group."
[143] Steve Calibro telephone interview by Richard Killblane, 14 February 2008.
[144] Calibro interview.
[145] The 1968 unit history report stated the 523rd had five gun trucks with a crew of four. The 1969 History Report mentioned their names: King Kong, Ace of Spades, Uncle Meat, Eve of Destruction, and True Grit. 1LT Robert M. Wills, Unit History, 523rd Transportation Company (Lt Trk), 1 January 1969 to 31 December 1969, 23 March 1970.
[146] James Lyles, The Hard Ride; Vietnam Gun Trucks, Quezon City, Philippines: Planet Art, 2002.
[147] 1LT Harris T. Johnson III, Unit History, 523rd Transportation Company (Lt Trk), 1 January 1968 to 31 December 1968, 31 January 1969.
[148] Ludy, "8th Transportation Group."
[149] Calibro interview.
[150] Ibid.
[151] Ibid.
[152] Ibid.
[153] Based upon the date on SGT Steve Calibro's award citation, and 1969 523rd Unit History Report.
[154] Calibro interview. Date and award verified in the 1969 523rd Unit History Report. The report also listed two other soldiers, SP4 James E. Pulley and PFC Jackie W. Witten, as having earned the ARCOM with V that same day. They were probably on Calibro's crew. 1969 523rd Unit History Report.
[155] Calibro interview; and Ludy, "8th Transportation Group."
[156] Calibro interview, Ludy, "8th Transportation Group;" and LTC William R. Saber, Operational Report of the 54th Transportation Battalion (Truck), WFR6AA, for Period Ending 30 April 1969, 5 May 1969.
[157] Calibro interview, and Ludy, "8th Transportation Group." The 1969 523rd Unit History listed five wounded in action during this ambush but not Calibro.
[158] 1969 523rd Unit History.
[159] Ludy, "8th Transportation Group;" and Saber, Operational Report.
[160] Ludy, "8th Transportation Group;" and Saber, Operational Report.
[161] James Lyles, The Hard Ride: Vietnam Gun Trucks; Part Two, n.p., 2011; and Calibro interview.
[162] Lyles, The Hard Ride.
[163] 1969 523rd Unit History.
[164] Joseph McCarthy II email to Richard Killblane, June 20-27, 2007.
[165] McCarthy interview; and Roger D. Champ interview by Richard Killblane, at Branson, MO, 18 June 2005.
[166] Champ interview.
[167] Terrance Van Brocklin telephone interview by Richard Killblane, 3 May 2011.
[168] Champ interview.
[169] Ibid.
[170] Van Brocklin interview.
[171] Ibid.
[172] Ibid.
[173] Ibid.
[174] Ibid.
[175] Ibid.
[176] Ibid.
[177] Champ interview, 18 June 2005 and telephone interview 1 September 2005.
[178] Champ telephone interview.
[179] Champ interview 18 June 2005 and telephone interview 1 September 2005.
[180] Ibid.
[181] Van Brocklin interview.
[182] Champ interview 18 June 2005 and telephone interview 1 September 2005.
[183] Joseph McCarthy email to Richard Killblane, June 27, 2007.
[184] Champ telephone interview.
[185] Army Navy/Vehicle Radio Communications (two way).

Vietnam War 75

QL19 looking west up An Khe Pass. Devil's Hairpin is on the top left corner before the road passed over the mountain.
Photo courtesy of the US Army Transportation Museum

Hairpin

Right before reaching the crest of An Khe Pass the incline steepened requiring the road to snake into a hairpin turn that slowed traffic to a crawl of about 4 mph whether climbing up hill or even going downhill and 2 mph negotiating the turn itself. After negotiating the hairpin turn the road straightened out and then curved to the left as it crested the Pass. Even after negotiating the hairpin convoys could not speed up for fear of creating too much of a gap between vehicles still crawling up the mountain side. Consequently, the convoys had to continue the slow speed above the Pass making them very vulnerable to attack going into the hairpin and coming out. For this reason the turn was called the Devil's Hairpin and Dead Man's Curve.[186]

A Republic of Korea (ROK) reinforced company had responsibility for the sector and established security check points throughout the pass. The next three examples show how the enemy tried to ambush convoys in that small area, all within a period of a couple of months.[187]

During 1969, COL Ludy had most of the mountain side defoliated leaving only scattered trees near the hairpin to deny the enemy concealment. The steep slope was strewn with boulders and by the next year patches of brush and elephant grass grew back that hid the enemy from observation making An Khe Pass a dangerous place for ambushes. In spite of the defoliation, the enemy still conducted ambushes.

Fortunately by January 1970, all task vehicles of the 8th Transportation Group had a sheet of armor plating placed in the window wall and under the driver's seat to protect them from small arms fire. This allowed drivers to remain in their trucks when halted in the kill zone and move immediately upon the direction of the convoy commander. Without the armor plating, the drivers had to exit their vehicles and take cover. This improvement enabled the trucks to exit the kill zone faster.[188]

During 1969, the enemy ambushed nine convoys in and around the Hairpin, and the next year ambushed five convoys in that same area. A convoy of the 54th Transportation Battalion was hit on 1 April 1970, and convoys of the 27th Transportation Battalion were ambushed on 10 January, 6 May, 10 and 15 June 1970. By this time, the convoy SOPs and gun truck tactics were well tested, so the following three ambushes illustrate the different ways the enemy used that terrain to ambush convoys.[189]

1 April 1970
54th Transportation Battalion

By 1970, the 523rd had six gun trucks; Eve of Destruction, King Kong, Uncle Meat, Ace of Spades, Black Widow and the Matchbox. The Matchbox began construction in January 1970 and was finished in February. It had armor plating 18 inches higher than average inside the bed of the 5-ton cargo truck and PSP inside armor. The gun box also had room for extra tires in the back. It initially mounted two M-60s in the front corners and a single M-2 .50 caliber machine gun center rear. It later ended up with three .50s, one forward behind the cab and two mounted on the side gun walls.[190]

On 1 April 1970, two westbound convoy serials headed up towards An Khe Pass. The gun trucks of the 512th escorted the lead serial that morning and Iron Butterfly was the rear gun truck of the first serial. It was the standard double walled 5-ton gun truck with a left rear side mounted .50 instead and two forward mounted M-60s. Matchbox and Uncle Meat of the 523rd escorted the second serial. They liked to work together, and the Matchbox liked to ride in the middle while Uncle Meat brought up the rear. The driver of Matchbox carried an M-79 grenade launcher and the crew had M-16s and fragmentary grenades. Uncle Meat also had three .50 cals, two front right and left and one rear pedestal mounted. Uncle Meat was also a maintenance gun truck with spare tires.

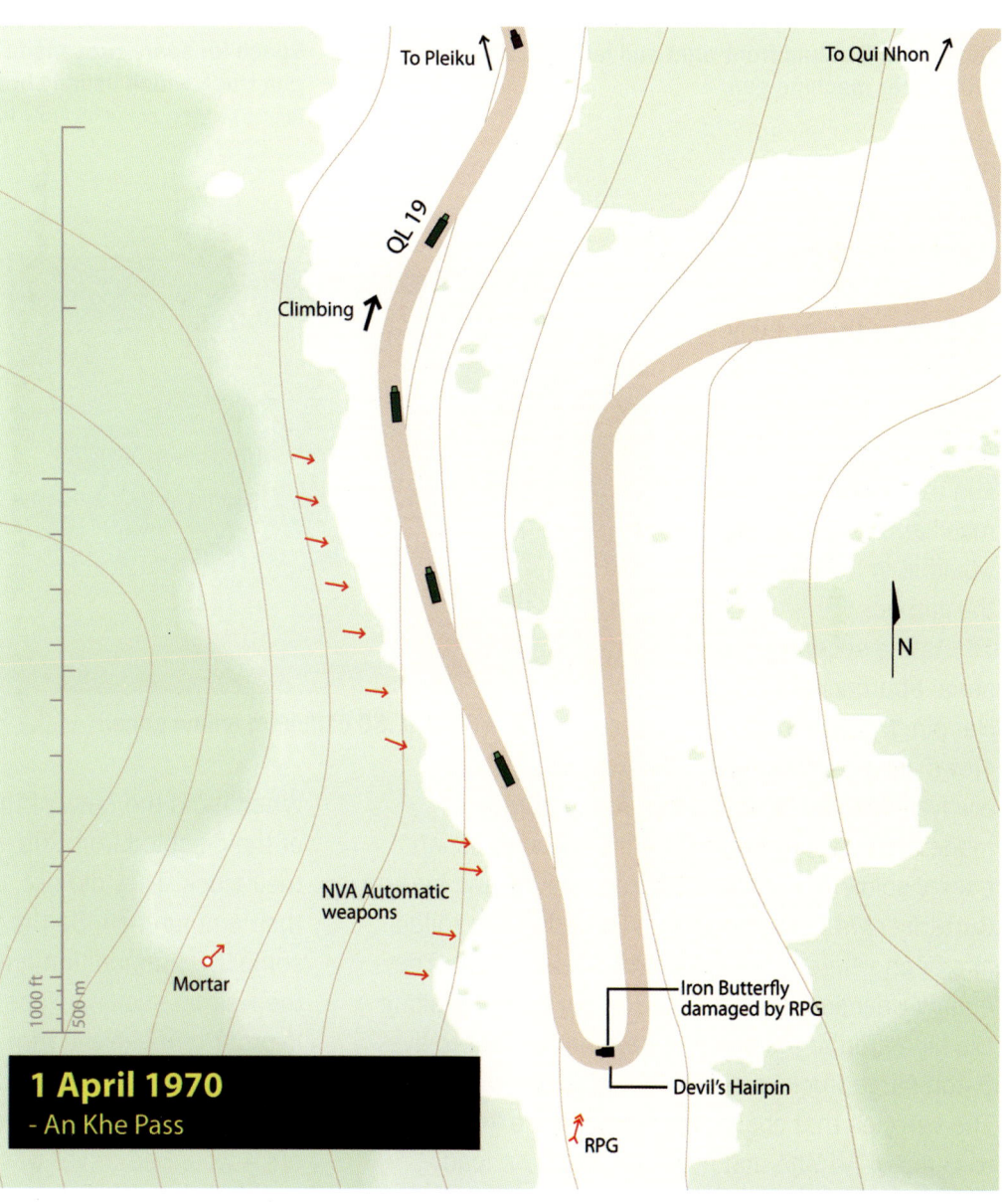

1 April 1970 - An Khe Pass

Vietnam War 77

The Match Box had front right and left .50s and one rear .50 caliber machine gun.

Space for spare tires made The Match Box a maintenance gun truck which usually brought up the rear of the convoy.
Photos courtesy of the US Army Transportation Museum

Both gun trucks had RT-524/VRC vehicle radios. The convoy commander drove up and down the convoy serial in his gun jeep to supervise the interval. All were standard 5-ton gun trucks and carried as much ammunition as would fit in the bed of the gun box.[191]

As the second convoy serial approached the Hairpin, the gun truck crews were alert as they always expected trouble there. As Iron Butterfly made the turn into the Hairpin in An Khe Pass, an enemy soldier directly ahead fired an RPG and hit Iron Butterfly right in the hood. Iron Butterfly then rolled into ditch where it remained and put up a fight. The enemy was directly above the Hairpin turn and along the high ground to its right. They shot up four to five 5-ton cargo trucks but only wounded one driver. The vehicles that could drove through the kill zone to camp at An Khe under the escort of the lead 512th gun truck.[192]

As soon they saw Iron Butterfly laying down heavy machine gun fire, the crew of Matchbox called in, "Contact, Contact, Contact!" Matchbox then drove up to assist the trucks left in the kill zone. Someone

Iron Butterfly with four .50 caliber machine guns.

made the decision for the trucks behind the kill zone to turn around and head back down to Cha Rang Valley. When some of the 5-ton tractors towing 105 howitzers on trailers turned around, they backed the tubes into the side of the mountain causing them to flip upside down. The Matchbox had a hard time dodging the trucks and bouncing 105s to get up into the kill zone. When it did the crew of the Matchbox asked the crew of the disabled Iron Butterfly if they wanted to leave their truck. They did not, but had fired off almost all their ammunition. So the Matchbox crew gave them some .50 ammunition and proceeded on up around the Hairpin sticking close to the left side of the road in order to make the turn as quickly as possible. As they cleared the Hairpin turn, three mortar rounds hit near them, but just over the edge of the drop-off. A wounded driver waved the Matchbox down. He had been peppered by shrapnel across his left side from the leg to his head by an RPG that hit the cargo bed just behind his cab. The gunners continued to exchange fire with the enemy while he climbed in the gun truck. The Matchbox then took him to the top of

78   Convoy Ambush Case Studies - Volume I

the Pass for a medevac then returned to clean up the ambush site. By then the enemy had broken contact and the gunners in Matchbox were completely out of ammunition.[193]

Uncle Meat drove up with the Korean reaction force walking in behind it. Uncle Meat then gave Iron Butterfly a push and it started. Iron Butterfly then drove up the Pass on its own power. Surprisingly, Koreans soldiers climbed out of the ditch where the ambush had taken place and used Uncle Meat for cover as they accompanied it up the Pass.[194]

View of the Hairpin from above.

Approaching the hairpin turn from below revealing the absence of trees. Photos courtesy of Ron Voithritter

Lesson
The enemy had initiated the ambush with an RPG on the gun truck in the rear of the convoy following by small arms and automatic fire. They had selected the Hairpin where the traffic had to slow down to 4 mph and had trouble not bunching up. The high ground above the road leading up over the Pass also allowed the enemy to shoot right down on the convoys and Larry Fiandt, driver of Matchbox during the ambush, always felt vulnerable making that turn.

Since the element of surprise allowed the enemy to destroy at least one vehicle, the NVA soon learned to take out the gun truck first. Dividing convoys into serials of 20 to 30 vehicles driving within supporting distance proved to be a wise decision. This allowed the Matchbox from the next serial to reinforce Iron Butterfly in the kill zone until Uncle Meat brought up the Korean infantry.

10 June 1970
54th Transportation Battalion

At 0845 hours on 10 June 1970, the lead element of a 17-vehicle westbound convoy reached the Hairpin just below the An Khe Pass and traffic slowed to a crawl. About 200 meters below the hairpin turn to the east, a low ridge extended southward that caused the road to curve at that location. The ground then rose slightly on the south side of the road.

The lead vehicle reached the Hairpin as the convoy crawled at about three to five miles per hour. An NVA soldier stood up on a slight rise to the south side of the road further back and opened fire with his AK-47 from approximately 25 meters away hitting the engine and shooting out several tires a refrigeration (reefer) van. Regardless, the driver continued to drive his rig for about 100 meters when another enemy soldier fired a B-40 rocket into the rear of his van. The driver still managed to drive his rig out of the kill zone. The convoy then came under enemy fire with small arms and automatic weapons from both sides of the road along a 200-meter kill zone and upon hearing the initial firing, the convoy commander led the first half of the convoy to the top of the pass. A gun truck and armored maintenance truck headed for the kill

Vietnam War

10 June 1970
- An Khe Pass

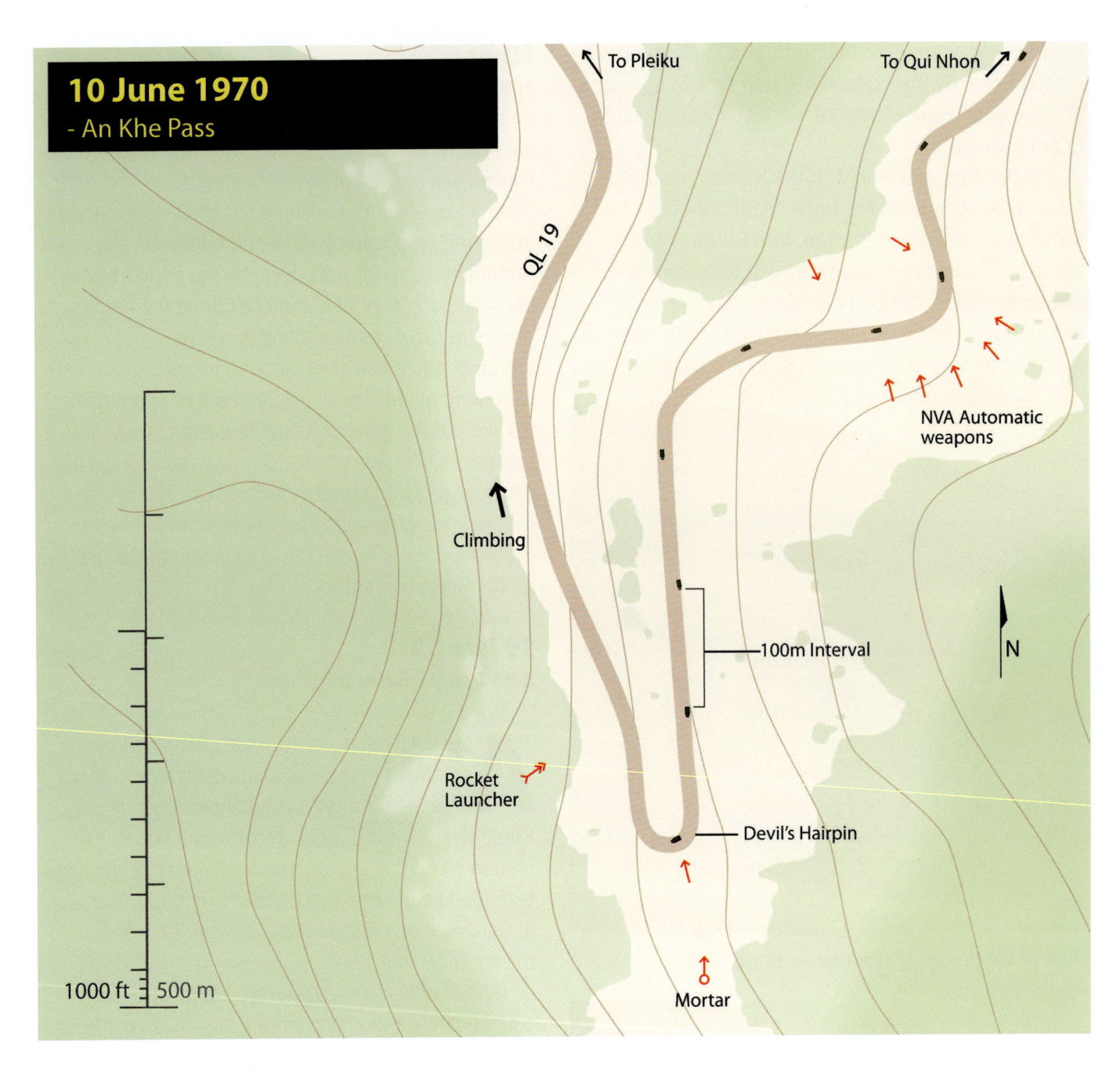

zone to lay down suppressive fire.[195]

The enemy had a heavy machine gun on a small knoll 200 meters south of the curve and the rest fired small arms and RPGs from 25 meters from the road. The enemy on a slight rise on the south side of the road poured heavy volume of small arms fire into the 40-foot Low Boy trailer, of the 444th Medium Truck Company, that was following directly behind the reefer van and disabled it. PFC Billy D. Wehunt, of the 444th Medium Truck Company, then drove his 5-ton tractor into the kill zone in an attempt to clear drivers out of the kill zone. As he approached the disabled Low Boy, the enemy fired into his cab killing him and his tractor drove off the side of the road.[196]

At the same time, a 444th gun truck drove into the kill zone to recover Wehunt from his vehicle, and came under intense small arms fire from above. Two MP V100 armored cars arrived and the gun truck crew again tried to recover the body of their fellow driver, but intense enemy fire forced the men back into their vehicles. Finally, two more gun trucks from the following convoy arrived and the combined fire power of five gun platforms provided enough covering fire for the recovery of PFC Wehunt's body. The enemy broke contact at 0945 hours, an hour after the ambush had started. The gun trucks checked all areas of the kill

80 Convoy Ambush Case Studies - Volume I

View of the kill zone above the hairpin where the reefer van was hit. Defoliation had killed off almost all the trees.
Photo courtesy of Ron Voithritter

zone for more personnel then proceeded to the top of the Pass. Helicopter gunships arrived after the convoy cleared the kill zone and worked the area. Korean soldiers conducted a sweep of the area after the gunships completed their runs. The Americans had lost one man killed, four wounded with three vehicles damaged.[197]

Lesson
The enemy selected a kill zone that provided elevation on both sides of the road where the convoy would slow down as the lead element entered the hairpin turn. The intensity of the small arms fire down into road on the north side took away the advantage of the steel wall of the gun box so the gun truck could not enter the kill zone without tremendous risk. It took the combined fire power of three gun trucks and two V100 armored cars to turn the fight back on the enemy or possibly the enemy just ran out of ammunition after an hour of fighting. The Koreans tried to deter further ambushes by establishing a prepared position near the hairpin curve.[198]

15 June 1970
27th Transportation Battalion

CPT Ronald Voightritter had been detailed to the Infantry for two years prior to changing to the Transportation Corps. He arrived in Qui Nhon in September 1969 and was initially assigned to the S-3 shop of the 27th Transportation Battalion. Around February 1970, he was given command of the 597th Transportation Company (Reefer). The 597th handled all the Reefer commitments for 8th Group and any tractor that did not have a reefer van hauled a regular cargo trailer. The company was short officers so he rode on convoys as the convoy commander about every three days and had never seen any contact prior to this ambush. The 597th had four gun trucks: King Cobra, Sir Charles, Poison Ivy, Blood Sweat and Tires.

15 June 1970, CPT Voightritter was assigned the convoy commander of a 39-vehicle convoy to An Khe and Pleiku. The task vehicles consisted of four lowboys, four reefers, seven roll-on/roll-off vehicles (commercial Sealand vans) off Landing Ships, Tank (LST), four Express vans with high priority cargo, and 19 flat beds from the 2nd, 444th and 597th Medium Truck Companies. He had no particular line up for the tucks other than the order they were lined up by the night shift at the marshalling yard. For escort he had two gun jeeps, the APC gun truck King Cobra with three .50s from his company and the 5-ton maintenance gun truck, The Saint, from the 444th. The Saint had a forward M-134 mini-gun, and two left and right front M-60s. He also had one wrecker and bobtail from his company. He placed the gun jeep with the NCO up front, King Cobra in the middle, The Saint in the rear behind a bobtail and the 2½-ton maintenance truck loaded with tires since it was a maintenance gun truck. As the convoy commander, Voightritter always rode in the rear from where he could drive up and down the convoy to control the convoy.[199]

Unfortunately, his regular gun jeep was with a convoy that had remained overnight at Pleiku and he was assigned gun jeep #66, which they kept off the road because it had maintenance problems. In fact, it broke down soon after the convoy cleared the gate;

so he went back and borrowed the S-3's jeep. It had no pedestal mount for the M-60, so Voightritter tossed the M-60 in the back seat with three to four cans of ammunition. His gun jeep only had him and the driver, SP4 Calvin Wood, both with their M-16s in the gun racks and an extra M-79 grenade launcher.[200]

Since Voightritter's regular assigned jeep and driver were up at Pleiku, he asked for Calvin Wood as his driver that day. Wood was the 3rd Platoon jeep driver. Voightritter had ridden with him several times in the past and considered him one of the best drivers in the company.[201]

Voightritter had called in the check points but heard no acknowledgement on the radio, which caused him to fear the radio on the S-3 jeep did not work. Prior to reaching the Hairpin, he drove up the convoy ensuring everyone had the proper 100-meter interval and their flak jackets and helmets on. He then fell back to the rear of the convoy. He was about 200 meters from the turn when he heard a boom directly above him followed by small arms fire. Thinking it was a mortar round, he immediately called, "Contact, Contact, Contact!" on the radio, then placed the M-79 in Wood's lap, and climbed into the back seat to fire the M-60. He then yelled at Wood, "Go," but Wood, who had been in several ambushes, had already hit the accelerator and raced into the kill zone propelling the captain into the back seat against the radio. As they came around the turn, Voightritter fired off one burst of the M-60 at the hill above him and the hot spent cartridges hit Wood's helmet and bounced down into the collar of his flak jacket causing him to yell in pain. Voightritter fired off three more bursts from the M-60 then it jammed so he climbed back into the passenger's seat, grabbed his M-16 from the rack

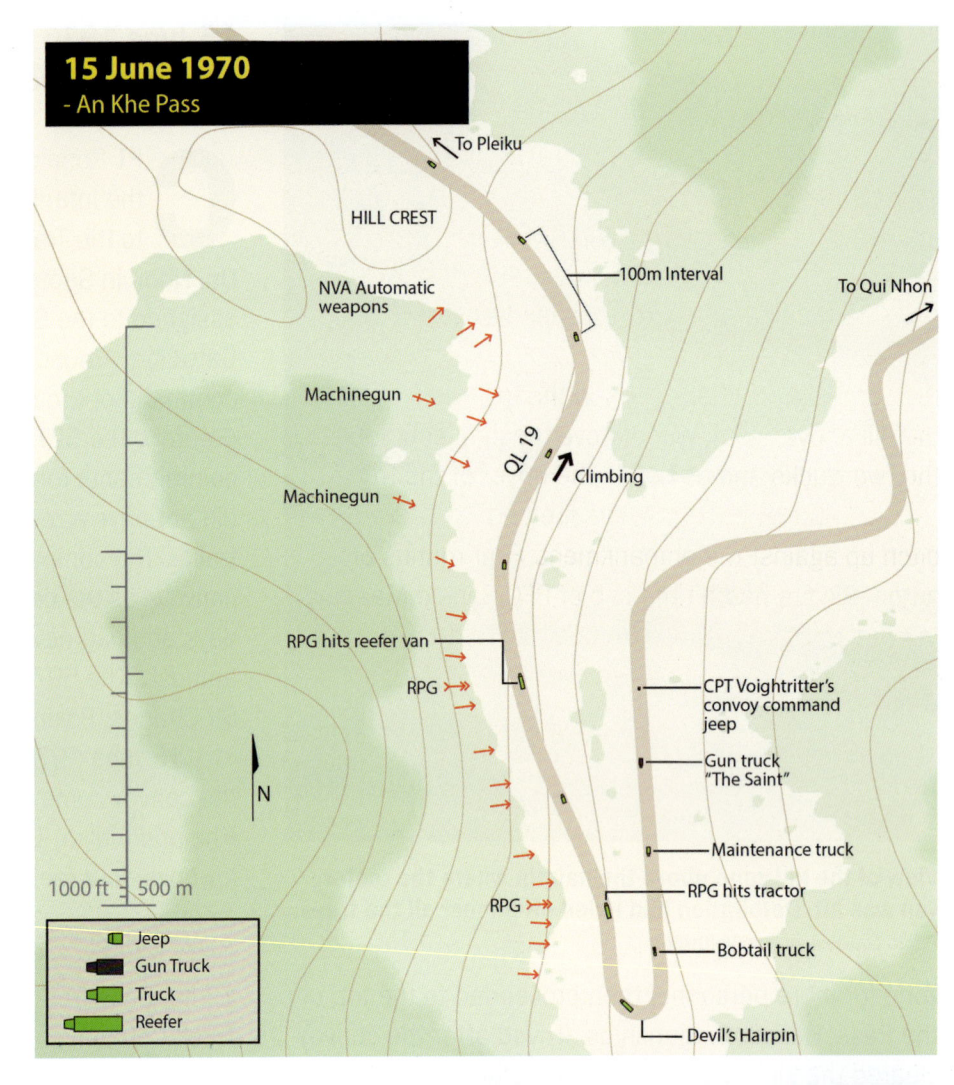

15 June 1970
- An Khe Pass

View showing gun and crew postitions on King Cobra.
Photo courtesy of the US Army Transportation Museum

82   Convoy Ambush Case Studies - Volume I

and squeezed the trigger. Voightritter had never been in contact before nor fired a round in his nine months in Vietnam, so there was no round in the chamber. Not sure of why the M-16 failed to fire, he then grabbed Wood's M-16 and fired it.[202]

The explosion had been an RPG that hit the second to last reefer and stopped it in the road. The last reefer was passing it when Wood and Voightritter rounded the turn. The Korean soldiers in the sandbagged check point in the turn were firing up the hill above them. Wood drove the gun jeep between the two trucks and as he did the driver of the truck passing was hit by small arms fire and went in the ditch up against the embankment. Four rounds of automatic fire had hit the cab of PFC Ronald Manning's truck; the armor plate in the door stopped three but one had come through the canvas top and hit him.

King Cobra and crew of the 597th Transportation Company. Photo courtesy of Ted Biggins

The Saint was maintence truck from the 444th Transportation Company. It brought up the rear of the convoy. Photo courtesy of James Lyles

Those that could drove up to the top of An Khe Pass. The Saint entered the kill zone and stopped next to the first disabled reefer while King Cobra was backing down into the kill zone both laying down suppressive fire.[203]

All the task vehicles made it to the top of the Pass except two trucks, so CPT Voightritter had Wood drive them back down into the kill zone looking for the drivers of the two trucks. Wood stopped his jeep behind Manning's rig in the ditch. While still under fire, Voightritter got out and ran to the passenger side door but it was locked, so he ran around to the driver's side and opened the door. Manning was slumped over and showed no signs of life. About that time, Wood climbed up over the armor plate and opened the passenger door from the inside. Voightritter then laid Manning down in the seat and slid him over so Wood could pull him out. They loaded him in the back seat of the jeep and then Voightritter ran down to the second truck. He slipped on the asphalt, slid down to the truck and grabbed the door handle. He then opened the door and saw no one was there. Wood yelled he had already checked the cab. They drove back up the pass to the check point at the bridge to evacuate the driver. When Wood had exited his jeep, he did not know the M-79 fell in the road so he left it. The infantry that swept the kill zone later found it. Driving they saw an M-16 in the road, so Voightritter told Wood to stop. Voightritter picked up the M-16 and fired it from the kneeling position in the passenger seat in his jeep. A helicopter happened to fly by and see the ambush. It then landed in front of the jeep at the top of the pass and they loaded up Manning for a flight to An Khe.[204]

While all this was going on at least one gun truck and one gun jeep from an eastbound convoy Voightritter's had passed at the bottom of the Pass had come up to help. The crew of the jeep was already talking to battalion so Voightritter used their radio to talk to battalion. The first thing Battalion said was, "Where to hell have you been?" Voightritter took this as an offense as if he had not been doing his job. He then explained he had one KIA and one MIA, and two gun trucks still in the pass. They told him to go back, find the missing driver, break contact and get the gun trucks out of the kill zone. He and Wood then

Vietnam War 83

made a third pass through the kill zone to get the gun trucks to break contact. At the top of the pass they recieved some additional fire from the VC but the gun trucks suppressed it.[205]

Voightritter then led his convoy to An Khe. At the next bridge he discovered one of his reefers. A rocket had ripped open the cab and it was hit by 11 automatic rounds wounding the driver in the leg, but he had driven to the top of the pass where a "shot gun" riding in an engineer low boy that had tagged alone with their convoy, got out and drove the wounded driver's rig up to the bridge. The infantry at the bridge treated the wounded driver and had called in a medevac for him.[206]

Coincidently, the NCO in the lead gun jeep had crossed over the top of the Pass and switched his radio frequency to the next security net before the ambush started so he had no knowledge of the ambush behind him. So the 88th Transportation Company sent gun trucks out of An Khe to chase down the front half of the convoy down and bring it back to An Khe. The convoy spent a couple of hours getting reorganized and taking care of things regarding the wounded. Voightritter found the missing driver was at the hospital with minor wounds. A rocket had disabled his truck and wounded him with shrapnel, so he jumped on one of the passing trucks. Manning, on the other hand, was pronounced dead and the medics asked Voightritter to identify the body.[207]

The convoy reorganized in the motor pool of the 88th Transportation Company and continued its mission to Pleiku. It received some mortar rounds in the Mang Giang Pass, which did not hit anything so the convoy kept moving on. It did not arrive at Pleiku until about 1500 hours so the drivers remained overnight (RON) there and returned to Phu Tai the next morning.[208]

Two vehicles were damaged with one soldier killed and three wounded in the ambush that lasted 35 minutes. The two damaged vehicles were left in the pass to be recovered later. It took place above the Hairpin where the last ambush occurred five days earlier. CPT Voightritter received the Silver Star Medal for his actions and SP4 Wood was awarded the Bronze Star Medal with V device.

84 Convoy Ambush Case Studies - Volume I

Top left: CPT Ron Voightritter, Commander of the 597th Transportation Company. *Left center:* SP4 Calvin Wood one of the best drivers in the company. *Left bottom:* Damage done to the reefer van by the RGP. *Top:* CPT Voightritter's gun jeep was on another convoy 15 June so he had to use another jeep with no pedestal mount for the M-60. *Above left:* View of last turn before the Hairpin showing QL 19 above turning back to cross over the mountain. *Above right:* This was the view CPT Voightritter would have seen of the kill zone above him.
Photos courtesy of Ron Voithritter

Lesson

Voightritter believed the armor plating of drivers' doors and the action of the gun trucks were a major life saver for 8th Group truckers. He felt the interval reduced the number of vehicles in the kill zone and limited the casualties. Because of the problem with the radio in the S-3's jeep, he felt he should have had redundant communications.

These three ambushes occurred in the vicinity of the Hairpin and the last two were within days of each other. The enemy hoped to take advantage of the slow traffic caused by negotiating the Hairpin.

Since it was difficult to concentrate a lot of fire power into the Hairpin, the key to survival was not bunching up. The convoy interval of 100 meters reduced the number of vehicles in the kill zone, so the enemy could only expect to damage or destroy a few vehicles at best. The enemy usually ambushed the middle or rear of the lead convoy. This split the convoy. The lead element automatically continued to safety with one gun truck as escort. That way, the enemy only had to contend with the rear gun truck, before the others from the next convoy could arrive. If the convoy commander was at the head of his convoy when the ambush split it, he was not in a position where the decisions needed to be made. The rear of the convoy increasingly became the location where critical decisions had to be made.

US Army Vietnam (USAV) published a monthly Combat Lessons Bulletin and in the October 1970 issue selected the 10 and 15 June ambushes for its issue on Convoy Operations. It identified the following lesson:

Both convoys were well organized and cleared the kill zone quickly. Gun trucks, and in the case of the second convoy, manned armored cars, moved in quickly to employ suppressive fires. Communications within the convoys and with supporting forces were excellent. Personnel in the convoys responded to the ambush with remarkable courage, professionalism,

CPT Voightritter and SP4 Wood receiving Silver Star Medal and Bronze Star Medal for their valor during the ambush.
Photo courtesy of Ron Voithritter

and responsiveness. In spite of the friendly counteraction, the enemy was able to inflict casualties to men and damage to materiel with a minimum cost to himself. This was primarily because the enemy gained the element of surprise. It is not possible to completely deny the enemy the element of surprise, but it is possible to reduce his opportunities to use surprise to his advantage.[209]

The convoy organization and procedures seemed sound and no one could see any change that would lessen the threat. So if the solution was not improving the convoy then they had to change something else. LTC Walter C. Daniels, the 27th Transportation Battalion Commander, summed up what his battalion learned from the three ambushes in his Operational Report of 13 August 1970:

During the report period, the enemy ambushed the battalion's convoys three times. All ambushes took place in the An Khe Pass in the area of the "hairpin curve". The enemy has an excellent vantage point in the "hairpin curve" area and is assisted in his endeavors by the fact the steep winding grade at this particular point has slowed the convoy's task vehicles to approximately 2 mph. In each ambush the enemy was believed to be in platoon to company size strengths and they fired small arms, automatic weapons, B-40 rockets and 60 mm mortars at the convoys. The battalion suffered 2 KIA's and 10 WIA's in the 3 actions. Following the last ambush, the battalion was given permission to burn the elephant grass in the area all three of the ambushes had taken place. The area was quickly burned to stubble and the enemy was no longer afforded the excellent concealment they had formerly enjoyed.[210]

The employment of convoy serials proved over and again the best method for mutual support. The gun trucks of the next serial could always seem to respond quicker than any combat arms reaction force. These ambushes set the stage for an ambush that would lead to the greatest act of courage among gun truck crews.

[186]"Ambushes in the An Khe Pass" is a study of the last four ambushes in the Pass written by an unknown author in 27th Transportation Battalion.
[187]"Ambushes in the An Khe Pass."
[188]COL Alex T. Langston, Jr., Operational Report of the 8th Transportation Group (Motor Transport) period ending 31 January 1970, 15 February 1970
[189]"Ambushes in the An Khe Pass."
[190]Larry Fiandt interview by Richard Killblane at Fort Eustis, VA, 14 September 2005.
[191]Fiandt interview, and James Lyles, T*he Hard Ride; Gun Trucks in Vietnam,* Part Two, n.p. 2011.
[192]Fiandt interview, and James Lyles, *The Hard Ride; Gun Trucks in Vietnam,* Part Two, n.p. 2011.
[192]Fiandt interview, and Lyles, *The Hard Ride,* Part Two.
[194]Fiandt interview.
[195]COL David H. Thomas, "Vehicle Convoy Security Operations in the Republic of Vietnam," Active Project No. ACG-78F, US Army Contact Team in Vietnam, APO San Francisco, CA 96384, 30 Sep 71; and "Ambushes in the An Khe Pass."
[196]"Ambushes in the An Khe Pass."
[197]Thomas, "Convoy Security Operations."
[198]"Ambushes in the An Khe Pass."
[199]Ronald Voightritter telephone interview by Richard Killblane, and 10, 12, and 15 May 2013; and Convoy Clearance # DW-166106, 15 June 1970.
[200]Voightritter interview.
[201]Voightritter interview.
[202]LTC (R) Ronald Voightritter email to Richard Killblane, 25 November 2003; and Voightritter interview.
[203]Ibid.
[204]Ibid.
[205]Ibid.
[206]Ibid.
[207]Ibid.
[208]Ibid.
[209]BG A. G. Hume, Deputy Chief of Staff (P&O), Combat Lessons Bulletin Number 14, Headquarters, US Army Vietnam, 20 October 1970.
[210]LTC Walter C. Daniels, Operational Report-Lessons Learned (Headquarters, 27th Transportation Battalion, Truck), for Period ending 31 July 1970, 13 August 1970.

359th Transportation Company (POL) Brutus, The Misfits, The Boss, Woom Doom and Ball of Confusion (later The Untouchable)

Of all the dangerous cargo to haul, fuel was the worst. It burned and every driver feared the slow painful death of burning to any other kind of death in combat. So the 359th Petroleum Truck Company was one of the more dangerous convoy units to serve in. SGT John Dodd was a career soldier and had begun his second tour in Vietnam in December 1968 where he was assigned to the 359th Petroleum Truck Company. The 359th had just transferred from Phu Tai to Pleiku in November 1968 and was attached to the 124th Transportation Battalion from the 240th Quartermaster Petroleum, Oil and Lubricant (POL)

Hauling fuel was the most dangerous cargo.

Battalion on 1 January 1969. Once the 240th Quartermaster POL Battalion had connected the pipeline all the way to Pleiku, this discontinued the need for fuel trucks to drive QL19. However, constant pilferage and interdiction by the enemy forced the Quartermaster battalion to shut down the pipeline. From then on fuel trucks had to drive the most deadly road in Vietnam.[211]

By the time Dodd arrived, the 359th had only constructed two gun trucks, Brutus and The Misfits, and a gun jeep. The new gun box design of the 359th was a smaller gun box of welded sheets of steel that fit inside the bed of the truck instead of the steel plates bolted on the outside. SGT Wayne Prescott helped build and became the first NCOIC of the Brutus, a 5-ton gun truck with two forward .50s and initially rear twin .50s. These were later replaced with a 7.62mm mini-gun in an armored gun box. The mini-gun's rate of fire instilled fear in the enemy. This six-barrel Gatling gun fired 7.62mm rounds at awesome speed, but was prone to misfiring. Dodd became the NCOIC of The Misfits, a 2½-ton gun truck with an M-60 and .50 caliber machine gun on a pedestal in the middle of the gun box. His crew consisted of a driver, John Hodges, and gunner, Bill Ward.[212]

Escorting fuel convoys with each tractor hauling 5,000 gallons of highly flammable jet fuel was probably the most dangerous mission for gun trucks. The enemy

359th POL convoy staging. Photos courtesy of 359th Transportation Company Vietnam

preferred to hit fuel tankers because the resulting fire usually blocked the road, trapped and destroyed the other trucks.

9 June 1969[213]

On 9 June, the Brutus, The Misfits and the gun jeep escorted a convoy of 30 fuel tankers on a return trip from Qui Nhon to An Khe. On the way down, SGT Barrowcliff had seen a uniformed VC carrying a LAW who escaped into the jungle. SSG Hutcherson rode in the lead gun jeep as the convoy commander with Roger Blink as his M-60 gunner and Jerry Usher as his driver. The Misfits drove in the middle of the convoy while Brutus brought up the rear. Peter Hish and Alan Wernstrum substituted in as gunners on The Misfits that day with Dodd, Hodges and Ward. Alan Wilson drove Brutus with Merton Barrowcliff on the M-60 and Prescott on the mini-gun. Barrowcliff was on his second tour in Vietnam. The convoy unfortunately had no air support that day. Once they reached An Khe, they would pick up the rest of the tankers from the 359th.[214]

Around 1600 hours,[215] the convoy had passed the Korean compound at the base of the An Khe Pass.

Barrowcliff standing in the right rear corner of Brutus looked down as it approached the last bridge and saw 20 to 30 VC lying along the road. Mert turned and screamed at Prescott, "Ambush!" then started for his M-60. Simultaneously, Dodd heard several rounds hit The Misfits' armor plating and heard Prescott, back on Brutus, scream over the radio, "Contact, Contact, Contact!" Dodd then spotted enemy movement about a hundred yards out in the field and returned fire. Barrowcliff looked to his left and saw the Korean compound getting mortared. From the trees, he then saw two rockets heading straight toward his gun truck. He then shoved Prescott's head down as the two rounds narrowly missed them and detonated on the other side of the gun truck. Barrowcliff again started firing his M-60, as the enemy walked mortar rounds down the road and two exploded in front of the truck ahead of his. Another landed right in front of his. He fired 2,000 rounds and then told Prescott to get off the radio and get on the mini-gun. In doing so he cleared a path. Meanwhile, The Misfits had cleared the kill zone and continued up An Khe Pass. Dodd then radioed back to Prescott and asked how he was doing. Prescott answered that Brutus had engaged about 40 enemy and the mini-gun was working fine. This was unusual as mini-guns were not designed for the road and the bumpy ride tended to knock out the timing mechanism. To keep the mini-gun operational, Prescott had to spend a lot of time working on it.[216]

As the road leveled out at the top of the Pass, The Misfits picked up speed again and led the convoy toward An Khe. It received some small arms fire but Dodd did not see anything to shoot at. The excitement passed as they left the danger behind them and the crew of The Misfits returned to their normal "chit chat" talking mostly about drinking a cold beer when they reached the security of An Khe. Dodd joked with the others while he kicked the .50 caliber brass around with his feet. The Dodd then called in the check point as The Misfits crossed a bridge about three miles from An Khe. In a few minutes they would be safe inside the security of the compound.[217]

A few seconds later he heard an explosion behind him followed by the sound of AK-47 fire. He turned to see where a mortar round had landed and the security force on the bridge was under fire. It was Dodd's turn to scream into the radio, "Contact, Contact, Contact!" He saw Viet Cong running around in the field to his left (south side of road) and opened fire with the .50 caliber. Hish and Ward worked as a team firing the M-60 while Wernstrum fired his M-16. Suddenly, someone on the radio asked for their location and size of the enemy force. This struck Dodd odd since he had just called in his location a few seconds before. Right after that an RPG slammed into the front portion of the gun box. The impact from the explosion knocked Dodd's feet out from under him but he did not let go of the machine gun. Wernstrum was lifting up another box of ammunition from the floor for the .50 caliber. Dodd then had Hodges pull the gun truck over and stop so they could provide suppressive fire while the rest of the tankers passed.[218]

A voice on the radio let him know that air support was on the way. Dodd was thinking short bursts with the .50 but his fingers called for long bursts. Suddenly a second RPG exploded on the gun box and blood splattered in Dodd's eyes. He looked down and realized he had been hit in the leg, chest and face but with the adrenaline pumping, he felt no pain. Looking around, he saw the blast had also blown Hish and Wernstrum out of the gun box and Ward lay on the floor clutching his stomach. Dodd realized in a flash all his gunners were wounded. He then called on the radio that he had three men badly wounded and needed a medevac.[219]

Climbing out of the cab and up on the corner of the gun box, Hodges checked on the crew. He then pointed to some water buffalo where he saw enemy movement. Dodd saw Viet Cong hiding behind the animals, which caused an elderly Vietnamese man to throw a fit knowing he was about to lose his work animals. Dodd told Hodges to get back in the cab and get ready to move out. Dodd then heard over the radio the medevac

Misfits on the back of a 2½-ton truck.
Photo courtesy of 359th Transportation Company Vietnam

Vietnam War 89

The Misfits had two twin .50s and side mounts for M60s.
Photos courtesy of 359th Transportation Company Vietnam

was on its way. He then picked up another box of ammunition and loaded it. Still conscious, Hish and Wernstrum crawled into the ditch on the side of the road, and Hish grabbed the tailgate to pull himself up when a third RPG hit the rear of the gun box knocking him back to the ground. Viet Cong then ran across the road and shot Hish twice as the M-60 tank and M-113 Armored Personnel Carrier (APC) drove up from the check point. The enemy then began engaging the tank and APC. This fire fortunately turned the enemy's attention away from overrunning the gun truck. Dodd realized he and Ward were wounded too badly to climb out and rescue the other two crew members. Ward needed immediate medical attention. Dodd slapped the top of the canvas with his hand signaling Hodges to drive off. When Hish saw the truck pull away he thought, "Oh hell, what am I gonna do now?" He then crawled back into the ditch. In half a minute The Misfits had cleared the kill zone and Dodd saw the medevac heading their way. By then Ward was sitting on an ammunition can clutching a one inch hole in his stomach. Dodd grabbed a large bandage and told Bill to hold it against the wound.[220]

Dodd then radioed back for Prescott to look for his two missing crew members. Prescott answered that the Brutus had tangled with some NVA after crossing the top of the An Khe Pass. The mini-gun had misfired but the crew managed to fight off an enemy rush on their gun truck.[221]

As his gun truck disappeared from view, Hish looked up and saw the welcome sight of a medevac helicopter. As it prepared to land, the pilot and crew saw the enemy dragging their wounded off into the jungle. They almost did not see Wernstrum.[222]

Down at the bottom of the pass, the convoy commander had driven back to the Brutus after it left the kill zone and asked the crew if they were okay. The crew signaled thumbs up. As they drove up the pass, the crew checked their guns and remained alert. About 300 or 400 yards from the hairpin curve, Barrowcliff looked up and saw around four unarmed NVA officers in the saddle on the west side of the road not 100 yards away. He started to shake since that was the third time in one day he had encountered the enemy. As Brutus approached, the enemy waved at them. Barrowcliff told Prescott who replied, "I don't believe it." He tried to call

Brutus was first built in 1969.

Road Runner on the radio but to no avail, then switched to An Khe Control and asked what was going on over the top of the pass because they heard a tank firing. Brutus then came upon three fuel tankers stopped on the right side of the road. The middle one was leaking fuel out of a hole in the right rear. The rear tractor also had blood in the cab. While still receiving small arms fire, Prescott searched for the drivers but could not find them. So he climbed in the leaking fuel tanker and pushed the truck in front out of the way, then drove it all the way to An Khe. Barrowcliff later asked him why he did that. He said, "It was DF2." When Barrowcliff told him it was highly flammable aviation gas, Prescott turned a little white and said, "It was not his day!" The battalion operations officer was mad at Barrowcliff because he had not found the two gunners blown out of The Misfits, but someone else picked them up.[223]

Meanwhile, The Misfits pulled into An Khe where the trucks had assembled. Hodges stopped the gun truck long enough to tell the drivers to let SSG Hutcherson know they had driven to the field hospital. As the gun truck drove 30 miles per hour though the gate of Camp Radcliffe, the MPs realized they were in trouble. Then two MPs jumped in their jeep and led them to the hospital. Dodd was looking after Ward's wounds when he saw that Hodges had the front bumper trailing just about two inches behind the lead jeep. An MP looked back and Dodd motioned for the MPs to either speed up or get out of the way.[224]

Once at the hospital, the medics helped the two wounded soldiers from the gun truck and put them onto tables in the receiving area. The nurses informed them both Hish and Wernstrum were already in X-ray and the medics then rushed Bill Ward straight to the X-ray room. The medics cut Dodd's clothes off and the doctor began pulling pieces of metal out of his legs. He informed Dodd the blood on his face came from the missing tip of his nose. Dodd's real concern was further down his anatomy. He kept trying to lift himself up to see what the doctor was doing to his legs, but the nurse kept pushing him back down. Because Dodd persisted in trying to rise up, she took her hands from his chest, then grabbed his "family jewels" and told him not to worry, everything is okay. Dodd fell back and relaxed.

While the doctors worked on Dodd, the medics brought in a wounded Viet Cong and laid him on the table next to him. In a loud voice, Dodd told them to move that SOB away a few more feet. Another nurse came in and told Dodd that his two missing crew members had been brought in. Pete Hish had fragmentary wounds and had been shot. Alan Wernstrum also suffered from fragmentary wounds and lost part of his hand. The four wounded crew members had a short reunion in the medical ward that evening and the EOD presented them the rocket section of an RPG that had hit their gun truck, which Dodd later donated to the Transportation Museum. Because of the seriousness of their wounds, Ward, Hish and Wernstrum were medevaced to Japan that evening.

Since the hospital at An Khe had become crowded, Dodd was flown to Pleiku where he hitchhiked back to his company area. He did light work supervising the Vietnamese labor around the company for about 30 days until his wounds healed and then he rode on South Vietnamese convoys calling in check points until he left Vietnam on 4 November 1969. SGT Prescott received the Silver Star Medal for his actions and Wilson the Bronze Star. Dodd received an Army Commendation Medal with V device but he threw it away.

Lesson
The enemy employed two kill zones, the first most likely hoped to draw off the gun trucks so the next ambush would catch the convoy of fuel tankers unprotected. Probably one of the boldest moves the enemy had made yet, in their effort at unpredictability, they had launched the main ambush right outside the compound at An Khe where most drivers felt relatively safe.

Radios in each of the gun trucks made them more responsive to enemy action and allowed them to act independent of the convoy commander. However, this placed greater decision making responsibility on the crews of the gun trucks.

The Misfits' gun box took three hits from RPGs, yet one gunner still continued to put up a fight. The enemy had to choose between a mobility kill which still allowed the gun truck to fight, but not maneuver, or go for the crew kill, which often failed because of the open design of the gun box. The open gun box allowed for the blast of an RPG to escape without killing the entire crew. The gun trucks could take great punishment and still continue to fight due to the dedication of the crews.

21 November 1970

By 1970, the 359th Medium Petroleum Truck Company had a total of five gun trucks; The Misfits, Brutus, Outlaws, Woom Doom and Ball of Confusion. The crew of Brutus consisted of William "Bill" Kagel, Ernest "Ernie" Quintana, and SGT Jimmy

Brutus initially had twin .50s on back and later they were replaced with a .50 amd a MI34 minigun.
Photos courtesy of 359th Transportation Company Vietnam

Vietnam War 91

Freeman firing mini-gun on The Untouchable.

Photo courtesy of Rumbo

Ray Callison. Brutus was still a 5-ton gun truck with two forward left and right M-2 .50s and a rear M-134 7.62mm mini-gun. The six-barreled M-134 mini-gun could fire at a rate of 4,000 rounds per minute.

On the morning of 21 November 1970, a jeep with radio led the 27th Battalion convoy serial of thickets bordered both sides of the road with the tree line 250 meters from the road.[225]

At 1105 hours, the middle of the convoy came under rocket, automatic and small-arms fire from the south side of the road. The 800-meter kill zone caught Brutus and six fuel tankers. The crew of Brutus called "Contact, Contact, Contact" over the radio and immediately returned fire with a .50 caliber and mini-gun. B-40 rockets ignited two of the tankers and another jackknifed partially blocking the road. Small-arms fire punctured the tanks and flattened the tires on the three other tankers but they managed to drive out of the kill zone, pick up the drivers of the burning tankers while leaving a trail of leaking fuel on the road. Brutus then joined the lead vehicles at the top of the pass and halted. The vehicles behind Brutus had halted, turned around and driven back down the road. Ball of Confusion and the convoy commander's jeep were two miles back down the road assisting a broken down vehicle. For the first few minutes of the ambush, Brutus bore the brunt of the fight.[226]

Upon hearing "Contact," the convoy commander and three gun trucks of the 597th Medium Truck in the eastbound convoy; Sir Charles, King Cobra, and Poison Ivy, turned around and raced back to the kill zone. Sir Charles was an APC gun truck with a forward .50 mounted on the track commander's hatch and two left and right rear M-60s on the troop hatch. King Cobra was also an APC gun truck but had three .50s. Poison Ivy was a standard 5-ton gun truck with a forward M-60, and right, left and rear .50 caliber machine

Photo courtesy of the US Army Transportation Museum

Photo courtesy of the US Army Transportation Museum

92 Convoy Ambush Case Studies - Volume I

guns. By the time they reached the kill zone, Ball of Confusion had preceded them. Because the jackknifed tanker blocked the road, the gun trucks bunched up on the east end of the kill zone and placed suppressive fire with all their weapons into the enemy positions. A rocket hit Brutus on the driver's side above the tire, wounding Kagel and Quintana and killing Callison. Ball of Confusion also had one man wounded during the ambush.[227]

This ambush was timed with an enemy attack on Landing Zone Attack, just three miles down the road. Within 15 minutes of the first call, six APCs and one

Jacknifed fuel tanker blocking the road. Brutus halted in the background.

Brutus halted with tanker blocking the road ahead.

A crater in front of disabled Brutus was most likely caused by a rocket.

Ball of Confusion halted two miles behind Brutus. Photos courtesy of Tim Sewell

Vietnam War 93

tank from the 1st Squadron, 10th ARVN Cavalry arrived in the kill zone. The majority of the ARVN Cavalry Squadron, which had responsibility for this area, had been pulled up to Pleiku several days before for operations in that area. After another five minutes of fighting, the enemy withdrew and only sporadic firing continued for another 15 minutes when two gunships arrived.[228]

Ronald Mallory, Richard Bond, Larry Dahl and Charles Huser had previously become friends with the crew of the Brutus. Every time the Brutus returned from a convoy, Mallory and his friends liked to help take the weapons off and clean them. They were curious and wanted to know everything about the gun truck. One day the crew told their friends, "If anything ever happens to us, we'd like you all to take over the gun truck." In that manner the crew of the Brutus had chosen their replacements. Ron Mallory, Chuck Huser, Larry Dahl and Richard Bond replaced the crew of Brutus. Because Mallory was one of the best drivers at split shifting, he naturally became the driver. Bond assumed the responsibility as the NCOIC with Huser and Dahl the gunners. As the new crew set about cleaning and repairing the gun truck, the loss of their friends saddened them. To give the truck a new look, they completely repainted Brutus. They thought this would make the old crew proud of what they had done. It took about a week to put Brutus back on the road, but took almost a month for the smell of blood to leave.[229]

Lesson

It was possible the enemy had hoped the attack on the landing zone would have drawn off the reaction force, however, the gun trucks held their own in this fight and the reaction force arrived. The gun trucks with their wild names and fancy artwork had added to the élan of the gun truck crews. This élan made them more independent of command and caused them to react without hesitation. The crews knew the key to turning the fight back on the enemy was to mass greater fire power against the enemy. As soon as the call, "contact," went out, all available gun trucks would rush into the kill zone. This would prove critical during the next ambush.

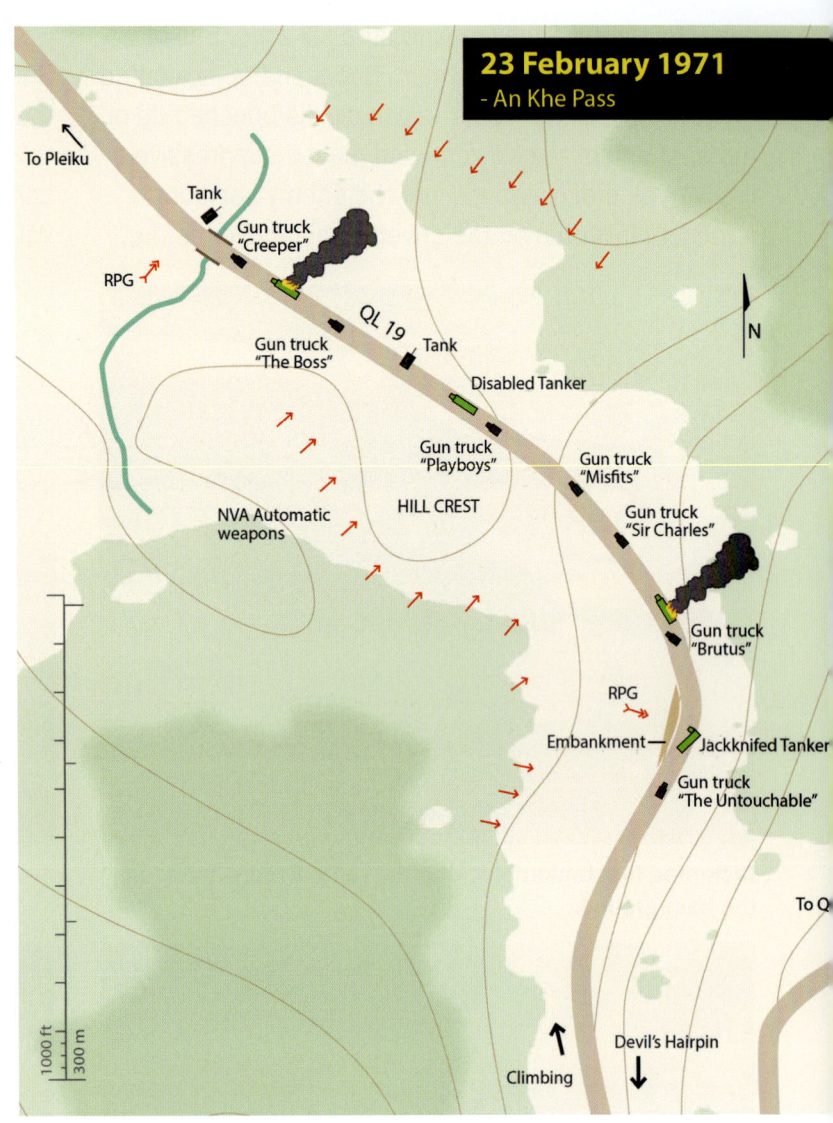

Dahl Ambush, 23 February 1971

On 23 February 1971, Creeper, The Boss and Playboys of the 545th Light Truck Company escorted a fuel convoy under the control of the 27th Transportation Battalion heading west to Pleiku. The convoy commander rode in a gun jeep. They left the staging area at the Ponderosa right after the roads were opened just after daybreak and operated under the call sign "Challenger." Creeper was the lead 5-ton gun truck with call sign, "Challenger Six." It had four corner mounted .50s so the forward and rear gunners could switch from left to right depending upon which side the attack came from. SGT Ellis "Mac" McEarchen was the NCOIC and operated the two forward .50s. Barry Montgomery was the rear gunner and William Mongle was the driver. Twin M-60s were mounted on

Photo courtesy of US Army Transportation Museum

a pedestal over the passenger side of the cab for the driver. Mongle was also armed with an M-79 grenade launcher and had a cut-down 55-gallon drum filled with 40mm rounds. The cab also had 1,600-round boxes of M-60 ammo. The gun box was a double wall design with steel outside and PSP on the inner wall holding double rows of sandbags. The inner wall was about ten inches lower than the outer wall and had a plywood lid over it with room to store extra M-60s, .50s and Thompson submachine guns. The spare .50s were worn out and required four clicks to adjust the timing, which made the barrels loose. The forward .50s had aviation barrels that would not overhead and require changing during a fight. The crew lined the bed of the gun box with two layers of 100-round ammo cans so they hopefully would not run out of .50 caliber ammunition. The rear of the gun box also had room for spare tires. When Creeper rolled down the road, the crew kept the barrels strapped down because the weight of the gun made it want to tip the barrel up. They had a quick release on the straps so they could quickly free the gun and place it in action.[230]

The Boss followed about seven to ten trucks behind Creeper. The Boss was a 5-ton gun truck with front right and left .50 caliber machine guns and one rear center mounted .50. As the NCOIC of the gun truck, SP5 John Chapman, was the forward gunner and operated the radio. Paul "Pineapple" Mauricio was the rear gunner. Walter Deeks was the regularly assigned driver but had the day off. Neither Paul nor John remembered who their replacement driver was that day, but as the driver he manned twin M-60s mounted on a pedestal over the passenger seat of the cab during an ambush. The Boss had the same double wall gun box design but no spare tires. It carried three M-79 grenade launchers stored in the wall compartment.[231]

Photo courtesy of James Lyles

Vietnam War 95

Photos courtesy of 359th Transportation Company Vietnam

The crews also had their M-16s and the driver was armed with an M-79 grenade launcher. Because he was a grenadier/driver on this convoy, Walter did not check out his M-16 that day. The standard basic load for all gun trucks was about eighty 100-round cans of .50 caliber ammunition, or about as many as would fit on the floor. 40mm grenade rounds in Playboys were stored in a footlocker between the cab and gun box.[232]

The 545th Light Truck had originally operated out of Tuy Hoa but recently moved north to Cha Rang Valley in December 1970. All the gun truck crews had experienced numerous small arms fire, mines and RPGs but no ambush on the intensity of the one they were about to drive into that day.

The second serial departed the staging area around 0800 hours with the 5-ton gun trucks, The Misfits, Brutus, and The Untouchable, and the gun jeep, Little Brutus, as escorts. The Misfits, call sign "Sugar Bear Two," led the convoy armed with two right and left forward .50s and pedestal mounted twin .50s in the rear. The crew that day consisted of Harold Shartle as the driver, Glen Kunston as the NCOIC, Kenneth Sheck as a gunner, and another gunner who was picked for the day but no one remembered.

Playboys, call sign "Challenger Eight," followed seven to ten tankers behind The Boss. The regular assigned driver of Playboys, Danny Cochran, had gone on sick call so the NCOIC, SGT Grailin Weeks, asked Walter Deeks to fill in for him. Walter did not like escorting fuel tankers because they tended to catch fire when hit by rockets; but since gun truck crews were considered the elite of truck drivers, he could not turn down the request. Gun truck crews did not turn down missions and many drove right up to the day they had to leave for the United States. Playboys was a 5-ton gun truck with a double wall design same as The Boss and armed with two forward and one rear .50 caliber machine gun and an M-60 over the cab. Garry Looney manned the rear .50, with Ronnie Louden on the right front .50 and SGT Weeks on the left gun so he could also operate the radio. Their lieutenant liked to ride in the last gun truck.

Brutus, call sign "Sugar Bear One," followed in the middle. It was a 5-ton gun truck with two forward right and left .50 caliber machine guns and a rear pedestal mounted 7.62mm mini-gun. Mini-guns put out a lot of firepower but required lots of maintenance to keep the timing right. SP4 Richard Bond had the day off to go before the E-5 promotion board, so SSG Hector J. Diaz filled in as the NCOIC of the Brutus that fateful day. Although SP4 Charles L. "Chuck" Huser was not the NCOIC, he operated the radio since he knew how his and the other gun trucks operated. Ronald "Ron" Mallory was the driver and SP4 Larry Dahl was one .50 gunner. The main crew had only served on Brutus for the last three

96 Convoy Ambush Case Studies - Volume I

Photo courtesy of Fred Carter

months. The Untouchable, call sign "Sugar Bear Three," followed behind it with SP5 Erik Freeman as the NCOIC.[233]

After the previous ambush that killed Callison, "Filthy Fred" Freeman replaced a wounded man on Ball of Confusion and became the NCOIC. He had reenlisted for a second tour in Vietnam just to be an NCOIC of a gun truck in the 8th Transportation Group. Freeman replaced the 2½-ton with a brand new 5-ton cargo truck and replaced the two M-60s with forward and rear mounted M-2 .50 caliber machine guns and left and right M-134 7.62mm mini-guns. Since mini-guns were not issued to truck units they were hard to get, so Freeman acquired enough damaged or worn-out parts until he could assemble two complete mini-guns and then turned them in for a direct exchange with an Ordnance unit for two brand new functional M-134 mini-guns. Because of the problems with timing, Freeman never let anyone else touch the mini-guns. Because he wanted to build the most powerful gun truck in Vietnam, he named his new gun truck, The Untouchable, after actor Robert Stack's popular TV series, "The Untouchables," that ran from 1959 to 1963. Freeman dropped the plurality of the name because he did not want the name to reflect the crew but instead the gun truck. The rest of the original crew was due to rotate home so Freeman recruited a whole new crew. He had no hesitation getting rid of anyone who did not measure up to his high standards and he did not tolerate drug use or heavy drinking. So his crew was very well disciplined. Baylon Braswell was the driver, and Robert Logan and John Ross were the .50 gunners. The Untouchable was a maintenance gun truck loaded with spare tires so it usually followed in the rear of the convoy.[234]

The convoy commander, LT Porter, rode in the armored gun jeep, Little Brutus, somewhere near the rear of the convoy. Little Brutus, call sign "Little Bear One," had armored doors and gun box with twin M-60s mounted on a pedestal. After he took over the gun jeep in January, the NCOIC and gunner, SP5 William Fred Carter, had replaced the single M-60 with twins because of the tendency of M-60s to jam. If that happened, one M60 would still fire. He had ten 100-round 7.62mm ammo cans, a LAW and several hand grenades. The driver, Edward "Ed" Bonner, was armed with his M-16 and an M-79 grenade launcher and had a .50 caliber ammo box full of 40mm rounds. Porter liked to ride behind either Brutus or The Untouchable. The radio had two headsets so both the convoy commander and NCOIC could listen to it, because SGT Carter did not trust green lieutenants.

The second convoy serial left around 0800 hours and traveled at 25 mph.[235]

Brutus in April 1971. Photo courtesy of 359th Transportation Company Vietnam

Vietnam War 97

Photo courtesy of US Army Transportation Museum

Photo courtesy of US Army Transportation Museum

As the lead convoy serial winded around Devil's Hairpin and approached the top of An Khe Pass it slowed down to a crawl, which stretched out the interval. The high ground sloping upward to the left side of the road was covered with brush while the right side had been defoliated. As it drove left around the second bend and crossed over the Pass, Creeper maintained the slow speed to allow the rest of the convoy catch up. A cutaway embankment dominated the south side of the bend. The level ground further down the left side of the road had scattered vegetation, but the wood line had been cut back between 200 and 300 meters to the right. A small bridge crossed a narrow stream about 300 meters ahead of the Pass. Up until then, convoys had been hit coming up the pass and not near the security check point.[236]

About 1045 hours, Creeper had just crossed over An Khe Pass where the ground leveled out and neared the bridge. Barry Montgomery in the back looked down to his left and saw what he thought were three ARVN soldiers. They suddenly stood up and pointed their weapons at his gun truck causing Barry to ask, "Mac, why are those ARVN pointing their weapons at us?" Mac quickly shouted, "Get down!" and then 7.62mm rounds started hitting the side of the gun truck. Barry then unhooked his .50s and fired back killing two or three of them. Suddenly the whole jungle 200 to 300 meters to their right opened up with what they estimated must have been an NVA battalion. It was the worst ambush they had ever been in and it scared the hell out of them. SGT McEarchen quickly called, "Contact, Contact, Contact! We need help, they shot my tires out!" The gun truck stopped and the crew returned fire. The tank on the other side of the bridge also opened fire and the concussion ruptured Barry's ear drums causing them to bleed. They were in the worst fight for their lives.[237]

Photo courtesy of 359th Transportation Company Vietnam

98   Convoy Ambush Case Studies - Volume I

The Boss was coming up the Pass when its crew heard "contact" over the radio. Chapman tapped his driver to enter the kill zone and advised Mauricio to get ready with the red smoke to mark the enemy position. They passed a disabled tanker and then several more. As the road leveled out to a field or rice paddy on the right flanked by jungle, a few grass huts dotted the hill to their left. As soon as The Boss approached, an RPG or mine exploded on the left side of the road. The gun truck then drew fire from that direction and the gunners saw dozens of villagers running for cover. They could see Creeper taking fire from the jungle to the right. Paul threw the red smoke on the left side of the gun truck and they immediately returned fire, Mauricio firing to the left and Chapman to the right. Chapman took one earphone off an ear so he could hear what was going on around him, but it was hard to hear the radio traffic with the other ear because of the loud booming sound of the .50s. The Boss came to a stop behind Creeper and a burning tanker, another gun truck came up from the west and stopped on the other side of the tank and bridge. About this time, an RPG hit the right front tire and fender of Creeper.[238]

SGT Weeks also heard McEarchin calling for help and told Deeks to get the truck in position to protect Creeper. Playboys likewise raced into the kill zone. The tankers not in the kill zone stopped and waited while Playboys drove past them. As Playboys rounded the embankment on the left side of the road right, an enemy soldier stood up in the ditch on the left to fire his B-40 rocket at the cab of Playboys. Walter Deeks stopped the gun truck and the rocket climbed passing overhead, while a .50 gunner fired right directly over Deeks' head killing the enemy soldier, but hurting Walter's ears. Playboys then proceeded around the bend into the kill zone passing a fuel tanker jackknifed on the right side of the road. The driver had evidently abandoned it. As Playboys came up on the level ground, Weeks and Deeks could see a 5,000-gallon fuel tanker on the right side of the road leaking fuel from bullet holes behind The Boss. Other tankers had evidently cleared the kill zone. They saw the immobilized Creeper a short distance ahead still placing suppressive fire on the enemy.[239]

SGT Weeks directed his gun truck pull up next to a disabled fuel truck on the right and then the crew placed suppressive fire on the enemy to the left that were very close to the road hiding behind rocks and even in the ditch. They were so close Weeks could see how young they were. After Looney fired off his first 100 rounds, he jerked the empty ammo can out of the feed tray and his steel pot fell off onto the road. He then loaded a second ammo can and something caught his eye. He turned and saw a truck driver sitting in the road about 50 yards behind Playboys waving at him. His legs were wounded so he could not get up. Looney hollered at Grailin and then pointed at the guy. Looney then climbed over the back of the truck, ran back to the wounded driver. The driver was a lot bigger than him so Looney grabbed the driver by his brown leather belt and dragged him behind the rear dual axel of the disabled truck.[240]

Looney then stood up and took two or three steps backwards to catch his breath, and then three NVA came from around the right side of truck. One pointed an AK47 at him. Looney sprinted as fast as could to his gun truck picking up his helmet on the way. When he got back up on the truck, Louden was on Grailin's .50 shooting to the left and Grailin was on Looney's rear .50 shooting to the right. So Looney got on the right front .50. About that time Deeks looked out through the armor plating and saw about 15 soldiers to his right either running away from the fire or moving to better cover in the wood line. Looney shot them and Grailin yelled, "You son of a bitch; those are ARVN!" Gary replied, "No, they're not. They're NVA! I saw them close up." So they continued to fire. The gunners did not fire three to six round bursts but employed spraying fire. Not knowing exactly where the enemy was, they depressed the butterfly triggers on the .50s and did not let up until they burned off a 100-round belt of ammunition. Weeks called for air support and handed Deeks two six-round bandoliers of 40mm grenades. The viciously intense exchange of gun fire continued for about 15 minutes, which seemed like an eternity to those in it, until the Cobras arrived and started tearing up the tree line.[241]

When the Cobras, call sign "Shamrock," arrived, they asked the gun trucks to mark their positions with smoke grenades and Looney popped purple smoke. The pilot warned them, "Y'all get your asses down. I'm coming to purple haze."[242] The gun ship fired mini-guns and rockets very close to Playboys on the high ground to the left side of the road. It made at least two strafing runs and then hovered and fired.[243]

Two columns of smoke rising from burning tankers in the kill zone. Photo courtesy of Timothy Sewell

The crew of The Boss did not wear their flak jackets or helmets and blazed away at the enemy. The gunners poured oil on their barrels after every 200 to 300 rounds to keep the barrels from overheating and Mauricio changed barrels only when his turned red, so he had to use the glove. He did not use a timing gage, but just spun the barrel in tight then backed off three clicks, fired and added another click until the weapon fired as it should. John Chapman on The Boss heard an eastbound convoy from An Khe ask if they needed any help and John told them, most likely Sir Charles from the 597th Transportation Company, to go to the next gun truck, Playboys. By the time the Cobras arrived, the right front .50 on Creeper had jammed but the gunners were so scared they forgot they had spare barrels in the wall compartment. After the Cobras arrived the enemy fire died down, but whenever there would be a lull in the fighting, more enemy soldiers would move to a better position and the fighting would intensify again. About half an hour into the ambush, everyone heard the approaching sound of mini-gun of Brutus and felt safer.[244]

The second serial was about a mile from the check point at the base of An Khe Pass when they heard over the radio, "Contact, contact, contact!" LT Porter then instructed the lead truck to throw a smoke grenade to mark its position and stop. Little Brutus then drove up the road directing the tankers to pull over to the side of the road. The drivers could see the two columns of smoke from the burning tankers rising at the top of the Pass. Someone on a gun truck asked over the radio if they should go help and Porter thought about it for a while. Little Brutus then pulled up and Porter got out to talk with the NCOICs of the gun trucks. He decided they would go help the convoy in the kill zone, and The Misfits, Brutus, The Untouchable, and Little Brutus drove up the Pass into harm's way.[245]

The Misfits cleared the bend and Brutus stopped halfway around the bend near a burning tanker and the embankment where the NVA soldier had fired the B-40 rocket at the Playboys. The Misfits, Brutus, and The Untouchable stopped about 100 yards apart. The ambush then had seven gun trucks (over 21 machine guns) and at least one tank engaging the enemy. SSG Diaz and Dahl fired their .50s to the right side of their gun truck while Huser fired the mini-gun. After a while, the mini-gun quit firing and Diaz and Huser quickly went to work to get it back in action. The bouncing of the road often threw the timing off. Freeman in The Untouchable could not see Brutus in front of him due to the hill between them but could hear everyone yelling on the radio. In their excitement, they forgot radio discipline. Suddenly, he heard, "They're on the left; they're on the

Brutus heading up An Khe Pass into the kill zone.
Photo courtesy of Tim Sewell

100 Convoy Ambush Case Studies - Volume I

left!" So Freeman moved to the left mini-gun. He could see a little bank that sloped uphill to a mound.[246]

About 15 minutes after the 359th gun trucks had entered the kill zone the fighting died down and someone called for a cease fire. Many assumed the fight was over while others scanned for targets. Freeman dismounted to talk with SGT McQuellen, the lead wrecker operator, and the lead convoy commander. After five minutes of the cease fire, someone told Ron Mallory, "Okay, we can go back and pick up our convoy. Turn around right here." Ron then backed Brutus up to turn around and suddenly three to five enemy soldiers charged Brutus from the high ground on the south side of the road.[247]

Diaz heard Dahl yell, "Frag!" An NVA soldier had lobbed a grenade into the Brutus. Diaz then turned and saw Dahl drop to his knees and cover the grenade with his body saving the lives of the rest of the crew. The explosion threw Dahl into Huser knocking him down. Shrapnel bounced off the inside walls of the gun box and hit Diaz in the back and legs also knocking him down. The other crews saw the explosion in the gun box and knew the crew was seriously hurt but not how bad. Ron Mallory heard the explosion looked down and saw he was covered with blood. He thought, "Lord, I've been hit." Since he had to drive them to safety, he had no idea what happened in the back.[248]

The crews of the gun trucks had developed a special bond of friendship. Many gun truck members expressed the same opinion that any crew member would have done the same thing to save the lives of their crew. Larry Dahl just saw the grenade first. Diaz saw Dahl try to get up and then fall back down into Huser's lap. Diaz asked how they were. Huser said Dahl was dead and he was wounded in the arm and legs. Huser then tried to call for help over the radio but it would not work. So he then yelled at Ron Mallory to drive them out of there. Brutus then made a hard left hand turn, backed up again and pushed the burning tanker out of the way, then raced his gun truck west through the smoke of the burning tanker to the nearest friendly check point at the bridge where a medevac awaited for the wounded.[249]

About that time the tank from the bridge drove over and stopped behind Playboys then fired its main gun two or three times. The blast ruptured Looney's ear drums. The enemy fire then died down again so someone told Deeks to get out and look for the driver of the disabled truck. A premonition told Deeks not to check the driver's door but to run around to the passenger's side. As he rounded the front of the truck, he saw a scared young NVA soldier staring up at him from under the wheel well. The kid did not look older than 14 years of age. Deeks did not have his M-79 with him and both he and the kid had a look of shock on their faces. So Deeks spun around and ran back the way he came shouting to the gunners, "There's one under that truck, there's one under that truck." Louden yelled, "There he goes." He then shot and killed the enemy soldier. Paul Mauricio on The Boss saw the fear on Walter's face and claimed he also blew the enemy soldier away. Deeks admitted he was shaking like a leaf.[250]

Burning tanker which Brutus pulled up to next to.
Photo courtesy of Tim Sewell

Someone yelled they saw the driver behind the truck. The blast had evidently blown him out of the truck. A tall, former high school basketball player, Deeks ran over and picked the wounded driver up in his arms. The bloody and unconscious driver was peppered with shrapnel. Deeks carried him over to the gun truck and the crew tossed down a first aid kit. He patched the worst wounds while the crew called for a medevac. Deeks told the gunners the wounded driver needed a stretcher, but the gun truck did not have one. They told him to ask the tank crew if they had one. Deeks felt terrified approaching the tank that was buttoned up and firing at the enemy. He jogged over while yelling at the crew not to shoot him. They said they did not have a stretcher so Deeks returned to his gun truck. After the helicopter landed,

Damage caused by shrapnel bounding off the inner wall.
Photo courtesy of Timothy Sewell

Freeman talking to LT Porter and SGT McQuellen with Brutus in the background next to the burning tanker seconds before it was attacked. Photo courtesy of Fred Carter

Deeks carried the driver over to it. Covered in the driver's blood, he returned to the cab of his gun truck and by that time the fight had quieted down again.[251]

About the same time Deeks went after the driver, SGT Weeks saw trucks trying to drive around the tanker directly behind Playboys, but it blocked the road. Without anyone telling him, he got out and climbed in the tractor. He did not remember the 5,000-gallon fuel tank was leaking fuel but Deeks did. They were still under fire and SGT Weeks knew someone had to get that tanker out of the way. He just focused on driving it off the road. He drove it into the field to the right, jumped out and let it come to a stop by itself. He then quickly climbed back in his gun truck. After that the ambush ended. The enemy had been dug in and seemed hell bent on destroying the convoy. The ambush had lasted for over an hour, a lot longer than the usual 10 to 15 minutes.[252]

About that time, The Untouchable and Little Brutus turned around and rejoined their convoy on bottom of An Khe Pass. The Misfits remained with Brutus at An Khe. Once at the bottom of the pass, The Untouchable and Little Brutus escorted their convoy all the way to Pleiku.

Meanwhile, The Boss at the top of the Pass pulled over to the left side of the road to let its convoy pass and fell back into its original position. Creeper had all but one tire shot out and had to roll on rims. The convoy passed the infantry on their way to mop up the ambush site. The convoy then stopped at the check point and the convoy commander told Deeks to get a count of trucks and drivers. He was still shaking from the ambush. Once

out of harm's way, they could relax a little. The shock of the event finally caught up with Deeks. As he was trying to get a count he fell on his knees and started vomiting. The other drivers finished the head count for him. Mauricio was so shook up he could not even drink a cup of water, but grateful drivers came up and thanked him for saving their lives. John Chapman remembered everyone was shaking, but Weeks began shaking uncontrollably and went into spasms. Weeks remembered he was scared like everyone else but when he jumped down out of the gun truck, he was hurting. The convoy commander called for a medevac to take him back to the hospital at An Khe where they sedated him. His crew picked him up on the return trip that afternoon. Freeman had lost his hearing for three days.[253]

Because the enemy targeted gun trucks, the truckers believed the enemy had bounties on them, and since Brutus had lost two crews in the last two ambushes, they believe it had the highest bounty. Grailin Weeks received the Silver Star Medal and the rest of the crew of the Playboys earned Bronze Stars with V device. The entire crew of Creeper was submitted for Silver Star Medals but received no awards. Larry Dahl was posthumously awarded the Medal of Honor. Dahl's Medal of Honor proved that even truck drivers could be heroes too.

Nearly everyone was shaken up from the ambush. Barry Montgomery got off the road and pulled details around the camp. After a month, he joined the crew of the gun truck, Bad Hombre. With most of his crew in the hospital, the loss of his friends caused Ron Mallory to quit driving and he worked in the motor pool until he rotated back to the United States. Hector Diaz went through extensive medical treatment and died in 2003 of Hepatitis, which he picked up in a blood transfusion. Chuck Huser returned from the hospital to join the crew of The Misfits. Walter Deeks stayed on The Boss until a home town friend was killed on Cold Sweat with 20 days left in country. This broke his heart so he got off the road ten days later when he had 30 days left in country. Fred Carter and John Chapman remained on their gun jeep and gun truck until a couple weeks out from going home. Paul Mauricio took Chapman's place as the NCOIC of The Boss until it was his time to return home. The crew of Playboys remained on it and survived two more ambushes. Freeman extended for six months to stay on The Untouchable but his first sergeant would not let him extend any more after that. Having to leave the gun truck he built broke his heart. Of all the ambushes and enemy contact they had encountered in Vietnam, none had ever experienced an ambush as long or as intense as the one on 23 February 1971.[254]

Lessons

The Size of the enemy forcer and the terrain lent itself to a Z- shaped kill zone with part of the force on the high ground on the south side of the road and the rest in the jungle on the north side of the road.

The enemy continued to initiate ambushes by attacking the gun trucks first. The enemy was either determined to destroy the remaining trucks in the kill zone but the quick arrival of The Boss and Playboys prevented them. The crew could not determine if the enemy was maneuvering to attack or escape. The arrival of the four additional gun trucks from the other serials turned the tide of the battle. Against a determined enemy, it took as many as five gun trucks with three machine guns each to beat back their assault. Even then, this was just enough to defend the disabled vehicles and rescue drivers. The determining factor in the ambush was fire power – the number of machine guns in the kill zone.

SP4 Larry Dahl.

One big difference from the convergence of gun trucks in the kill zone the year before, this time the gun trucks maintained their 50 - to 100 - meter interval between gun trucks. This interval, unfortunately, placed Brutus next to the embankment where the enemy would attack.

The death of Larry Dahl taught everyone the importance of never letting their guard down even when it seemed like the ambush was over. Fred Carter warned, "Never drop your guard at any time. Always expect the unexpected, and when in convoy, always be in the defensive mode of operation."

Action Taken

After that ambush, 7th Squadron, 17th Air Cavalry was required to low and slow over the convoys. At first they flew around the country side but that did not deter the enemy from ambushing the convoys. The helicopters were directed to fly low and slow along the roads looking for wires and enemy spider holes. They flew so low that the drivers could reach up and touch the skids. This was boring duty for the pilots but they challenged their flying skills by landing on the back of moving trailers or flying between the trucks in the convoy.[255]

Vietnam War 103

Recovery Mission, 16 December 1970

On 16 December 1970, in a westbound convoy, Satan's Chariot passed a broken-down tractor and trailer and two gun trucks from an earlier convoy at An Khe Pass. Satan's Chariot was a 5-ton gun truck with three .50s, one forward and two rear. The convoy arrived at the 54th Transportation Battalion base camp just an hour before dark. SGT Charles Sims, the NCOIC of Satan's Chariot, had his men take off the weapons and start cleaning them when the commander of the 88th Light Truck arrived and told him to have his gun truck escort a spare tractor back up to retrieve the disabled vehicle. Sims challenged the decision. At that hour they would reach the pass after dark. Nothing traveled on the roads at night. He felt the gun truck should just pick up the driver, abandon the truck, and return, but the battalion commander issued Sims a direct order to recover the vehicle. The commander knew that if the tractor was left unattended on the road overnight was the Explosive Ordinance Demolitions (EOD) men would have to clear the vehicle the next day before a convoy could pass. Any vehicle left unattended on the road overnight was automatically considered mined. This would delay the departure of the convoys by an hour. Sims departed with a wrecker and spare tractor.[256]

When they arrived, Sims saw that the MP V-100 armored car that closed down the road each night was providing security for the broken-down tractor. Sir Charles also arrived. Once they recovered the tractor and trailer, Satan's Chariot led the way followed by the wrecker towing the tractor and trailer, Sir Charles and then the armored car. When they reached the base of the mountain, they found that the Koreans had strung a concertina barricade across the road closing Bridge Number 8. Sims radioed their situation to the road controller. The controller, in turn, called the American liaison officer with the Koreans to have them open the bridges. After waiting 20 minutes, the Koreans received instructions to open the bridge.[257]

Sims had driven the road so many times he thought he could have done it blindfolded, but it did not look the same in the dark. He did not remember the small village located on the north side of the road near the next bridge. Since there were no lights, he did not have visual cues to remind him of his location. To Sims' surprise, the next bridge was also closed. They waited for another ten minutes for the Koreans to open the barricades but these delays made them a target for a hasty ambush. They slowly negotiated around the barricades and all but the last gun truck and armored car had crossed the bridge when an explosion hit Sir Charles. The convoy then started receiving small-arms fire from the village on the north side of the road. Somebody screamed over the radio, "Contact!" Then the gun trucks, the armored car, and the Koreans immediately returned fire.

Air cover proved to be the best deterrent against ambushes. Photo courtesy of US Army Transportation Museum

Escorting convoys was boring so pilots would fly along the road in convoys like trucks or even land in the back of empty flatbed trailers. Photo courtesy of US Army Transportation Museum

Earlier the maintenance gun truck, The Untouchable, had towed a vehicle back to Cha Rang Valley. On his way back to camp, SGT Erik Freeman passed Satan's Chariot and the wrecker heading back up to the Pass. Freeman became concerned that two trucks up on An Khe Pass might have to remain there over night. He had a policy that he would never leave any of the vehicles or drivers behind if at all possible.[258]

Upon dropping off the tractor, Freeman's crew changed a flat and refueled while he monitored the radio. He could hear the recovery team was having trouble hooking up the broken-down tractor and worried the Koreans would close the bridges on them when it became dark. Fearing his commanding officer would not let him back out on the road at night, Freeman and his crew loaded up in The Untouchable and left without permission. According to policy, MPs at the gate were not allowed to stop a gun truck. Fortunately, The Untouchable found all the bridges open except Bridge Number 7. The Koreans let The Untouchable onto the bridge but would not let it cross. When Freeman saw the recovery convoy approaching, he convinced the Koreans to let his truck drive off the west end of the bridge to turn around. The Untouchable backed up and turned around then pulled up alongside the bridge so it would be ready to fall in with the convoy when it passed. Just as the wrecker crossed the bridge, The Untouchable pulled into the convoy and then small-arms fire broke out from the village lasting about a minute. The rocket blast had mortally wounded the NCOIC of Sir Charles.[259]

When the ambush started, the Koreans quickly closed all the bridge barricades. Freeman had just told his driver to back up when he heard the ambush, but The Untouchable was trapped on the bridge with Sir Charles and the armored car behind the bridge. Satan's Chariot could not reenter the bridge, so Sims led the wrecker back to Cha Rang. The rest of the bridges were open and he received a flare ship to escort his convoy back. Meanwhile, one of the gunners on The Untouchable pointed his .50 caliber machine gun at the Koreans, forcing them to open the bridge's west end. The Untouchable then turned around and pulled up next to Sir Charles. Freeman then raked the tree line with the mini gun for about 10 to 15 seconds and the other vehicles got under way and crossed the bridge. A medevac picked up the NCOIC, but he was already dead. An attack helicopter and flare ship then escorted Freeman's convoy back. This verified that any prolonged halt made vehicles a target of opportunity since local Viet Cong lived in the area.[260]

A local Vietnamese who lived in the village was killed in cross fire so the next day, the Vietnamese had painted a message on the road claiming that the Americans had killed her. This became an international incident, but the chain of command backed their gun truck crews.

Lesson
This was a hasty ambush by a small enemy force that took advantage of the delay in recovering the broken down tractor. The enemy also liked to fire on the Americans from villages so that some of the noncombatant villagers might be killed by the return fire. They could use this as part of their propaganda.

Well intended plans resulted in a series of delays in recovery and for whatever reason; there was a serious break down in coordination with the Koreans on opening the bridges after dark. Murphy's Law trapped these vehicles out on QL19 after dark, a dangerous place to be. In spite of these mistakes, the dedication of the gun truck crews to not leave anyone behind brought several gun trucks to the aid of one disabled tractor and trailer. In an ambush, the side with the superior fire power usually wins.

[211] Jon Dodd wrote up an account of the 9 June 1969 ambush and brought it to the 2004 Gathering. 1LT Alvin L. Preble, Unit History, 359th Transportation Company (Medium Truck Petroleum), 1 January to 31 December 1968, 6 April 1969.
[212] Dodd, Story.
[213] LTC John C. Kramer, Operational Report 124th Transportation Battalion (Truck) for period Ending 31 July 1969, 8 August 1969 (RCS CSFOR-65) (R-1) stated the date of the ambush was 9 June while Dodd's documentation stated 8 June.
[214] Dodd, Story.
[215] 124th History Report recorded the ambush taking place between Bridges 15 and 16 between 1530 and 1630 hours.
[216] Merton Barrowcliff, Story unpublished account sent to the 359th Transportation Company Vietnam, and Dodd, Story.
[217] Ibid.
[218] Ibid.
[219] Ibid.
[220] Ibid.
[221] Ibid.
[222] Ibid.
[223] Barrowcliff, Story.
[224] This and the next two paragraphs taken from Dodd's Story.
[225] Ronald Mallory interview by Richard Killblane at Ft Eustis, VA, 14 June 2004 and Thomas, "Vehicle Convoy Security," p. B-1-4.
[226] Thomas, "Vehicle Convoy Security," p. B-1-4.
[227] Thomas, "Vehicle Convoy Security," p. B-1-4, Mallory and Freeman interviews.
[228] Thomas, "Vehicle Convoy Security," p. B-1-4.
[229] Mallory interview.
[230] Barry Montgomery telephone interview by Richard Killblane, 4 September 2013.
[231] John Chapman telephone interview by Richard Killblane, 5 September 2013; and Paul Mauricio telephone interview by Richard Killblane, 5 September 2013.
[232] Walter Deeks interview, July 2000 and 29 March 2013; and Grailn Weeks telephone interview by Richard Killblane, 10 October 2013.
[233] Fred Carter had just stood before the promotion board the week before, which was why he remembered Bond had the day off to go before it that day. Fred Carter telephone interview by Richard Killblane, 20 September 2013; Gary Looney telephone interview by Richard Killblane, 25 February 2014.
[234] Erik Freeman interview by Richard Killblane at New Orleans, LA, 11 July 2002.
[235] Carter interview.
[236] Erik Freeman interview by Richard Killblane at New Orleans, LA, 11 July 2002.
[237] Deeks remembered hearing, "Help, help, help," Chapman remembered hearing, "ambush, ambush, ambush," but Fred Carter remembered hearing, "Contact, contact, contact!" followed by, "We need help," and then something about the gun truck being disabled. Montgomery interview.
[238] Mauricio and Chapman interviews.
[239] Deeks interview, July 2000 and 29 March 2013; and Weeks interview.
[240] Deeks interview, July 2000 and 29 March 2013; and Weeks interview.
[241] Looney interview, and Weeks interview.
[242] With a little play on words, "Purple Haze" happened to be a 1967 song by Jimmy Hendrix.
[243] Weeks interview.
[244] Montgomery, Chapman, and Mauricio interviews.
[245] Fred Carter telephone interview by Richard Killblane, 20 September 2013. SSG Hector J. Diaz statement after the ambush said the gun truck, The Untouchable bolted past us and the burning tanker entering the kill-zone." Erik Freeman, NCOIC of The Untouchable remembered he stopped behind Brutus, but a photo taken by Fred Carter, showed Freeman talking with the lead convoy commander and wrecker operator with Brutus behind them.
[246] Deeks interview, Freeman interview; and Diaz Statement, no date.
[247] Not everyone was in agreement how soon after its arrival was Dahl killed. Diaz Statement claimed 15 minutes after its arrival. Fred Carter remembered the fighting had died down after they arrived, Deeks and Montgomery remembered hearing the sound of the mini-guns firing when Brutus arrived in the kill zone. Most agreed the entire ambush lasted an hour or more, so 45 minutes into the ambush seemed a reasonable estimate of when Dahl was killed. Carter and Mallory interviews; and Diaz Statement.
[248] Freeman, Deeks, and Mallory interviews; and Diaz Statement.
[249] There are differing opinions on which direction Ron drove. Fred Carter remembered Mallory drove Brutus back down the pass rather than try to negotiate the kill zone. The gun trucks crews ahead remembered Brutus driving past them in the kill zone heading west. Mallory said he was in shock but thinks he drove westward to the nearest check point manned by an Infantry unit at the bridge. He remembered driving through the smoke of the burning tanker. Diaz Statement also recorded they drove to Bridge 19, which was past the bridge where Creeper was hit.
[250] Deeks, Chapman and Mauricio interviews. Mauricio remembered this incident took place ten minutes after their arrival, but Deeks remembered it took place after Brutus was attacked.
[251] Deeks interview.
[252] Deeks interview, and Weeks interview. Mauricio remembered Chapman told him at the end of the ambush it had lasted an hour and 25 minutes. Over 40 years later, Chapman remembered it lasted over an hour.
[253] Chapman, Deeks, Mauricio, Montgomery, and Weeks interviews.
[254] Carter, Chapman, Deeks, Freeman , Mallory, Mauricio, Montgomery and Weeks interviews.
[255] Paul Blosser interview by Richard Killblane, 9 June 2002 and LTC Alvin C. Ellis, "Operational Report – Lessons Learned, 39th Transportation Battalion (Truck), Period Ending 30 April 1971, RCS CSFOR-65 (R2)," Headquarters, 39th Transportation Battalion (Truck), APO San Francisco 96308, 4 May 1971.
[256] Email correspondence between Erik Freeman and Charles Sims, 15 – 21 February 1998.
[257] Ibid.
[258] Freeman-Sims email and Freeman interview.
[259] Freeman interview.
[260] Freeman-Sims email, and Freeman interview.

Ban Me Thout Pass. Photo courtesy of Wayne Patrick

2. Southern II Corps Tactical Zone
500th Transportation Group, Cam Ranh Bay

In spite of the losses incurred in the northern II Corps Tactical Zone by 8th Group convoys in 1967 and 1968, the 500th Group down in Cam Ranh Bay had not lost anyone to an ambush until late 1969, almost two years after the ambushes began along QL19. The threat level was not nearly as dangerous as it was on QL19 although enemy attacks on convoys in the southern II Corps Zone began to increase in the summer and fall of 1968. Of the two truck battalions under the 500th Group, the 24th Transportation Battalion conducted port clearance and local haul in the Cam Ranh Bay area while the 36th Transportation Battalion conducted line haul to the different camps. To accomplish this, the 36th Transportation Battalion had four truck companies: the 172nd, 442nd, 566th and the 670th Medium Truck Companies. The 172th was an Army Reserve Company from Nebraska that served in Vietnam from 1968 to 1969. Just as the 8th Group prior to COL Ludy, the 36th Transportation Battalion ran with long consolidated convoys of 80 to 150 vehicles divided into serials of 20 to 30 vehicles with a 5 to 10 minute gap between them and its gun truck development followed the same process of experimentation.

Between August and October 1968, the 36th Transportation Battalion converted 5-ton cargo trucks to hardened trucks just as the 8th Transportation Group had the previous year. Initially, the 5-ton cargo gun trucks added the cab armor kits and constructed double wall gun boxes out of lumber with the air gap filled with sandbags, with three gunners in the gun box. Each gun truck had two pedestal-mounted M-60s and one M-79 grenade launcher with a basic load of two 400-round cases of M-60 ammunition and one case of 40mm grenades. Like the 8th Group experience, these wooden gun boxes turned out to be too heavy resulting in poor handling, excessive wear on the tires and continuous brake failures and resulted in the death of one driver. To decrease the weight and still provide best protection, the battalion discovered a ⅜-inch rolled steel plate set two to three inches in front of a ¼-steel plate worked better. This reduced the weight to less than four and a half short tons. The battalion initially placed one gun truck in each convoy. The 36th Transportation Battalion

The Mortician. The first design of the guntruck in 500th Transportation Group were contructed of either 2 X 8 or 2 X 10 inch wood walls reinforced with two layers of sandbags. Photo courtesy of James Bambenk

Rear view of Ejaculator. Photo courtesy of James Bambenk

Above: The Flying Dutchman was one more plank taller. Photo courtesy of James Bambenk

Left: Front view of The Flying Dutchman. Photo courtesy of James Bambenk

Below: The 360th Transportation Company slid APC hulls in backwards to better distrute the weight for stability. Photo by Tom Berry courtesy of James Lyles

Vietnam War 109

learned the hardened trucks deterred enemy ambushes and after January 1969 increased the number of gun trucks from one to two per convoy, but recommended increasing the number of hardened gun trucks to six per medium truck company so each super convoy would have six gun trucks.[261]

Gun truck training in the 566th was learned on the job. Initially the gunners test fired their M-60s the night before the convoy at the post range, but often the gunner who fired the M-60 the night before was not assigned to the gun truck the next day. So the 36th Battalion wanted a test fire range next to the convoy assembly area.[262] Crews learned fields of fire and how to control fire while ensuring that the convoy drove through the kill zone. SGT Rich Cahill, who joined the 566th in January 1969, developed the tactic of having the gun truck in contact pull out to the contact side to make room for the convoy to pass then wait for the next gun truck to come up and take its place. Then the first gun truck would escort the convoy out of the kill zone. The M-2 .50 caliber machine guns did not arrive until around March 1969 and the crews initially mounted one in each gun truck. If the barrels of the .50 had to be changed out in the fight, the crews used the timing gauge to ensure that the new barrel had proper headspace even under fire. One unwritten SOP was that whichever side the convoy received contact, the gunner on the unengaged side would become an ammo man. Each sergeant ran his gun truck differently but typically when engaged, the driver stopped and fired the M-79 grenade launcher, the NCOIC fired the .50, another gunner fired the M-60, and the other gunner would serve as the ammo man. The rules of engagement still only allowed the gunners to return fire when fired upon and the mission of the 566th gun trucks was not to attack the Viet Cong but to defend the convoy to prevent causalities.[263]

In April 1969, SGT Rich Cahill's gun truck #363 followed right behind the lead gun jeep a large convoy heading up to Ban Me Thout. The road followed a northerly path up the ridge on the east side of the pass with the low ground on the west side. Like Mang Giang and An Khe Passes, the enemy liked to ambush convoys heading up this pass. On this mission, Cahill's convoy was hit by small arms fire from the woods. SGT Cahill drove up there to provide suppressive fire while the convoy raced past him. When the second gun truck came up, it laid down suppressive fire and Cahill's gun truck was supposed to pull out and keep going. Instead it remained in place because enemy fire came from one small area and Cahill wanted to make sure the rear elements caught up since they had the maintenance personnel.[264]

In late April 1969, a convoy from the 36th Battalion was bound for the 2nd Squadron, 1st Cavalry base camp at Phan Rang south along the coastal highway (QL1). This was a short run that only took half a day to reach the destination and then return by the end of the day. The convoys ran with 70 to 80 trucks with three to five gun trucks. They usually had an MP gun jeep or V100 armored car in the lead and gun trucks spaced evenly throughout the convoy with one in the trail party. The line up consisted of the MP gun jeep, the convoy commander's gun jeep, Rick "Snuffy" Smith's gun jeep, then the cargo trucks.[265]

The coastal highway, QL1, ran north and south along the flat coastal plain. The "Coconut Grove," was about halfway between Cam Ranh Bay and Phan Rang to the south. It was a rubber plantation on the west side of the road and a field of elephant grass on the east. Smith claimed that a million monkeys must have lived in the grove and swarmed the trucks whenever they passed, which was why they called it the "Coconut Grove." The coastal highway was a heavily trafficked highway by both military and civilian traffic. For this reason, no one ever expected to get hit along this route. So the drivers did not even wear their flak vests.[266]

When the convoy reached the "Coconut Grove," the enemy initiated the ambush with small arms fire on the lead and middle vehicles creating two kill zones. The APC gun truck, "USA," and one gun jeep were caught off guard. There was one gun truck and one gun jeep for every 30 vehicles. The vehicles were typically bunched up with no more than 20 feet between vehicles. The lead kill zone caught 16 to 20 vehicles. Gun trucks did their normal routine. Rick Smith fired his M-79 grenade launcher from the hip as fast as could. The enemy fired a few mortars but missed because they overshot. The fire fight lasted about 15 minutes.[267]

In spite of the number of trucks shot up, no one in the convoy was killed or wounded. Most of the trucks had flat tires, but all were able to drive to their destination. Rick Smith learned never to take his eyes off Coconut Grove again. The lesson was to expect an ambush where one least expected it. The enemy watched the convoy behavior and looked for signs of weakness.[268]

Meanwhile, through February and April, the 500th Transportation Group tested the concept of using the V100 Commando armored car as a security and control vehicle and liked it, especially because of the shortage of steel. It had better protection and fire power than the gun jeeps. The V100 had twin .30 caliber machine guns in a turret, which was far superior to an M-60 mounted in the back of an M-151. The V100 could attain the same speed as the M-151. So by November 1969, all convoys of the 36th Battalion were escorted by internal hardened gun trucks and MP V100s.[269]

In May of 1969, the 566th switched over from wooden gun boxes to steel. The US Navy provided sheets of steel in addition to the gun box kits. Gun boxes were made of 5/8-inch steel, wood inside the steel box, and two layers of sandbags for protection. By then, the crews of the 566th began naming their gun trucks. Jim Darby received The Mortician form the 172nd, Cliff Taylor named his The Gambler, and Walter Robertson named his Cobra's Den. The first gun truck to construct a steel box was The Mortician. Eventually the other five gun trucks received steel boxes, the Gambler and Cobra's Den being the last two.

The 360th Transportation Company (POL), although belonging to the Quartermaster battalion, rode in the 36th Battalion convoys. The 360th Medium Truck Company had built three gun trucks: USA in the 1st Platoon, Roach Coach in the 2nd Platoon and Grogin's Heroes in the 3rd Platoon. The 442nd Medium Truck built the Flying Dutchman, Widow Maker and Ejaculator.[270] The one difference between 8th Group APC gun trucks and those of the 36th Battalion; the Cam Ranh Bay crews slide their APC hulls in backwards which more evenly distributed the weight across the front and rear wheels. Even by January 1970, the three companies in the battalion had not replaced all the wood and sandbag gun trucks with steel boxes on account of the shortage of 5/8-inch steel sheets. The battalion also requested radios for each gun truck.[271]

The 36th Transportation Battalion did not have enough men to man all the machine guns in the gun truck, so LTC Edward Honor, 36th Transportation Battalion Commander from July through December 1969, circumvented that problem by borrowing fill-in gunners and volunteers from Finance, Admin, supply depot, and even Navy units around Cam Ranh Bay. Thus, only the NCOIC and driver were actually permanent crew members of the gun truck. The temporary gunners were taught to fire the machine guns on the range before going out on one convoy and then returned to their regular duties.[272] John Jacobs was one of the exceptions. He arrived in the 557th Maintenance Company in December 1969 and after seeing the gun trucks tear up a hill side at the range a week later, he signed up as a gunner and rode with SGT Cahill. After that Jacobs was hooked and signed up every day he returned to go out the next day. He rode a couple times with the 360th and 442nd, but mostly with the 566th. Otherwise, these temporary gunners provided no consistency.[273]

24 November 1969
36th Transportation Battalion

Ban Me Thuot was a routine destination for convoys of the 36th Battalion. The terrain between Nha Trang and Ban Me Thuot was mountainous jungle with some open areas where the jungle had been cleared or defoliated. The road was so narrow that trucks could not pass. The run to Ban Me Thuot took most of the day so the convoys usually had to rest over night at the camp and return the next day. The convoys to Ban Me Thout ran with anywhere from 80 to 150 vehicles divided into serials of 20 to 30 vehicles with a 5 to 10 minute gap between them. An MP with a V100 armored car usually led each serial followed by a lieutenant or NCO in a gun jeep with radio communications and an M-60 mounted on a pedestal. Each serial had a gun truck with an NCO, radio and one .50 caliber machine gun and two M-60s. The gun trucks carried twelve 100-round cans of M-60 ammunition and a like number of .50 caliber ammunition. The trail party made up the last serial and included a gun jeep, wrecker, ambulance, tire truck, 10 to 20 bob tails and a gun truck. The number of extra bob tails depended on the size of the convoy. Because of the rapid promotion from second lieutenant to captain in two years, LTC Honor had a policy that captains had to be the convoy commanders.[274]

Ban Me Thuot Pass. Photo courtesy of Wayne Patrick

On 24 November 1969, CPT Wayne Patrick, Commander of the 442nd Medium Truck Company, was the convoy commander. On a return trip from Ban Me Thuot, the convoy was delayed on account of the poor weather conditions and low clouds prevented helicopters from flying. Normally the convoys departed between 0700 and 0800 hours depending on mechanical problems. After line-up for the return trip they waited around an hour or more for the weather to improve. LTC Honor also had a policy that no convoy would run without air cover. It would have been normal procedure to radio battalion headquarters and inform them of the situation then get approval. Air cover would normally have joined them before they had gone far outside Ban Me Thuot. CPT Patrick made the decision to depart without air cover since road security in the mountain pass was considered adequate.[275]

The 101st Airborne Division and ROK Army provided security in the area. The Koreans had a base at the top of the Pass and the ARVN had a training base at the bottom. CPT Patrick had radio contact with security operations when entering their area of operation and there had been no reports of any significant enemy activity. It was not unusual to receive sporadic small arms fire from time to time but no convoy had been ambushed on this route before. Another factor in making the decision to depart was to return to Cam Ranh Bay before dark. It was not unusual to delay departure for various reasons but it was unusual to cancel a return trip.[276]

An MP V100 armored car and a gun jeep led the convoy. The convoy had between 80 and 100 vehicles divided into serials. CPT Patrick kept a gun truck at the rear of the first serial, another in the trail party, and the others space evenly though the middle of the convoy. He often rode either in the rear of the first serial or the middle of the convoy. This day he rode in the middle. This allowed him to drive up and down the convoy to respond better to problems. It also kept him in radio range with the lead and rear of his convoy. As the convoy commander, CPT Patrick's jeep had three radios to coordinate with air, ground and artillery support.[277]

Ban Me Thuot Pass. The enemy usually fired down on passing convoys with small arms from the high ground and crew-served weapons from across the valley.
Photo courtesy of Wayne Patrick

112 Convoy Ambush Case Studies - Volume I

Convoys to Ban Me Thuot usually waited at the bottom of the pass for air support. Photo courtesy of Wayne Patrick

About two hours after departure, the convoy was halfway down the mountain pass in the area secured by the 101st Airborne Division. The mountain rose above them on the north side of the road, to their left, and leveled out into a flat cleared zone to the south (passenger side) with a tree line around 100 yards away. That section of paved road had multiple curves that caused the trucks to slow down. CPT Patrick heard a boom up ahead followed by the report of "Contact" on the radio.[278]

The enemy in the wood line fired three to five B-40 rockets (RPGs) at one of the gun trucks in the middle of the convoy and hit the top corner of the passenger side of gun box. The blast wounded three crew members and cut SP4 Larry E. Collins in half. The lead part of the convoy continued on while the trucks behind stopped. Another gun truck pulled security on the disabled gun truck. The fight lasted five minutes.[279]

CPT Patrick was a quarter to a half a mile behind it and raced up to the rear of the disabled gun truck. He reached the scene a few minutes later. He was on the radio with the 101st. A couple of 101st paratroopers were nearby and were firing on the tree line for another 15 to 20 minutes. In just a few minutes there was a call on his radio from a Cobra gunship, with the call sign "Undertaker." He reported his position and three Cobras came in and worked the area over for about ten minutes. During that time, a Huey from the 101st Airborne Division came in and extracted the wounded. The rest of the convoy continued on. The trail party recovered the disabled gun truck.[280]

The rest of the convoy moved up while the area was secure and the trail party secured the disabled gun truck. The rear half of the convoy regrouped with the lead half at the normal rest stop in a safe area. The convoy returned to Cam Ranh Bay without further incident.[281]

Peacemaker waiting to escort a convoy of the 442nd Transportation Company. Photo courtesy of Wayne Patrick

Vietnam War 113

Lessons

Wayne Patrick drew these lessons: It was standard operating procedure to have air cover. Although we had sporadic enemy contact when air cover was in the area it was a deterrent to enemy activity. Radio communications within the convoy and with area security forces were extremely important. Quick response from area security forces limited the enemy's ability to continue with the attack. The Convoy Commander needs to be in the middle of the convoy to be able to respond quickly. The convoy maintained security and cleared the kill zone as per normal instructions. Clearing the contact/kill zone was always covered in briefings with convoy officers and NCO's. Drivers and others on the convoy knew to clear contact/kill zone.[282]

Patrick added these lessons:
(1) Have air cover up or very close by
(2) Have security forces on notice for quick response, make radio contact
(3) The Convoy Commander should have more than one radio, one to run the convoy and one for air support or security forces.
(4) You should have multiple gun trucks or armored vehicles per convoy in order to protect and secure the convoy in the event of attack or multiple attacks.
(5) Briefings should be held prior to departing to ensure that all personnel are aware of critical instructions to follow in the event of an attack.
(6) Make sure your weapons work prior to departure, test fire your weapons. This wasn't a problem for the 442.[283]

On 17 January 1970, COL Hank DelMar, Commander of Cam Ranh Bay Support Command, zeroed out the Headquarters and Headquarters Detachment, 36th Transportation Battalion and attached its truck companies to the 24th Transportation Battalion, which picked up the line haul mission in addition to the port clearance and short haul.

26 June 1970
360th Transportation Company (POL)

A normal line haul mission to Ban Me Thuot had 80 to 120 vehicles. Ronald Smith remembered the convoy on 26 June 1970 had 12 gun trucks; four from the 360th POL Company, four from the 670th Medium Truck Company and four from the 442nd Medium Truck. The reefers were in front, followed by the flat beds with projectiles (projos) and food, then POL tankers and

Photo courtesy of Wayne Patrick

the trail party. Convoys were organized by type of truck then by company. This would have placed the 360th POL tankers in the last march unit of the convoy.[284]

SSG Jack Buckwalter was the NCOIC of the 360th march unit. It started out with 21 vehicles. The platoon leader was the convoy serial commander and rode in a gun jeep at the head of the serial with a MP V100 and a gun truck. Another gun truck rode closer to the rear. A gun truck and three more gun jeeps were even distributed through the serial. Cliff Taylor's gun truck, The Gambler from the 566th, brought up the rear of that march unit. Walter Robertson's Cobra's Den also of the 566th was the rear gun truck and SSG Buckwalter rode in the trail gun jeep at the end of the convoy. This was the policy of 500th Group convoys at the time. Ten 10-ton tractor and trailers loaded with petaprime were waiting for them at the bottom of the Ban Me Thuot Pass so they could have air cover going up the Pass. The convoy had an L-19 Bird Dog for air cover. The ten tractors and trailers fell in the last serial.[285]

On its way up the Ban Me Thuot Pass, an enemy soldier fired an RPG at one of the lead POL tankers driven by SP4 Charles Pedigo. The rocket flew at an angle through the cab and hit the fuel tank. Pedigo safely jumped out of the cab but the truck started to roll forward. So he jumped back in the cab to set the hand brake then the fuel tanker blew up killing him. That act prevented the truck from rolling back down the steep grade into the other trucks. The burning truck melted to the asphalt and blocked the narrow mountain road stopping the convoy.[286]

The convoy commander and the vehicles ahead of the burning fuel tanker continued to Engineer Hill at the top of the Pass. He radioed back to his NCOIC that they were receiving small arms fire. They received small arms fire from the ridge across the valley to the north and down from the ridge above to the south. The enemy on the ridge across the valley had 81mm mortars and 12.7mm machine guns firing green tracers. All the gun trucks entered the kill zone, pulled off to the side of the road and fired at the enemy on the opposite ridge. The air cover left because there was too much ground fire. One gun truck pulled up to right next to the burning tanker, the USA pulled over to the right side of the road about 300 meters below the burning tanker returned fire.[287]

SSG Buckwalter only heard the muffled explosion of the rocket. He did not hear the small arms fires that far back in the convoy. He immediately raced his gun jeep to the front of the convoy. As he rounded the jungle road, he saw two MP gun jeeps halted on the left side of the road. He ordered them to drive up to the fight but they refused. Buckwater told his gunner to shoot them if they did not move. The MPs then did as instructed. When the SSG reached the scene, there was no longer any enemy fire. He told the crews in the gun trucks to open fire anyway on the ridge across the valley. During the process, a major kept calling on the radio wanting to know about his petaprime.[288]

The gun trucks returned fire for 35 minutes until the firing quieted down. SSG Buckwalter then made the decision for the trucks behind Pedigo's burning truck had to turn around and return down the mountain. The mountain road was narrow so the gun truck USA had to back up to a place where the tractors could turn around. Smith could see the smoke for 35 miles. They found an American unit at the base of the hill and waited for two and a half hours until the lieutenant called that they had air cover again. The drivers did not like the idea of driving back up the hill but Buckwalter made them. On the trip up, he stopped to recover Pedigo's body from the burning truck. The body had been burnt into a small ball and a few bones. He could not find the skull. The next day he stopped again but none of the body remained. Around 200 yards past the burned out tanker, the convoy was hit again by small arms fire but the gun trucks returned fire and the convoy kept going. The last gun truck dropped back to make sure that SSG Buckwalter was safe. It drove just ahead of the trial gun jeep the rest of the trip down the mountain.[289]

Lesson

This was one time that air cover did not discourage the enemy from attacking. The enemy attacked from the ridge across the valley. SSG Buckwalter thought the initial RPG round was fired from the gully below the road. That was why it hit its target. The small arms fire may have been ineffective because of the range. Convoys had been hit several times in this location and the enemy appeared to have tried attacking from different locations.

Since the enemy destroyed a fuel truck in the front part of the convoy, only the convoy commander and few vehicles were able to continue to safety. Fortunately, the most experienced leader was in the rear. SSG Buckwalter had already completed a couple tours in Vietnam. He was probably the most experienced soldier in the convoy serial. In this case placing the NCOIC in the rear of the convoy and the lieutenant at the head was a good idea. From atop the Pass, the lieutenant was able to coordinate the air cover for the convoy to return.

---

[261] LTC Paul E. Reise, Operational Report of the 36th Transportation Battalion (Truck) for Period Ending 17 October 1968, 31 October 1968, RCS CSFOR-65 (RI). MAJ Jonathan R. Barrett, Operational Report of the 36th Transportation Battalion (Truck) for Period Ending 31 January 1969, 12 February 1969, RCS CSFOR-65 (RI).
[262] Reise, Operational Report of the 36th.
[263] Interview by Richard Killblane of Clifford Wesley Taylor, Jim Darby, Ron Owens, Walter Robertson, Richard Cahill, John Jacobs, Harold Marshall, Alford Hatton, and Manny Tellez, veterans of the 566th Transportation Company at Pigeon Forge, TN, 7 August 2009. Barrett, Operational Report of the 36th.
[264] 566th TC veteran interviews.
[265] Rick Smith telephone interview by Richard Killblane at Branson, MO, 22 August 2005.
[266] Ronald Smith interview by Richard Killblane at Branson, MO, 17 June 2005; and Rick Smith interviews.
[267] Ronald Smith and Rick Smith interviews.
[268] Rick Smith Interview.
[269] 1LT James F. Hunt, Operational Report – Lessons Learned, Headquarters, 500th Transportation Group, Period Ending 30 April 1969, 5 September 1969; MAJ John F. Nantroup, Operational Report of the 500th Transportation Group (Motor Transport) for Period ending 31 July 1969, RCS CSFOR-65 (RI), 8 December 1969; and LTC Edward Honor, Operational Report – Lessons Learned, Headquarters, 36th Transportation Battalion, Period Ending 31 October 1969, 25 February 1970.
[270] Ronald Smith interview; and Wayne Patrick interview by Richard Killblane at New Orleans, LA, 12 July 2002.
[271] MAJ Thomas P. Storey, Operational Report – Lessons Learned, Headquarters, 36th Transportation Battalion, Period Ending 17 January 1970. Headquarters, 36th Transportation Battalion (Truck) APO 96312, 17 January 1970.
[272] Richard Killblane, Mentoring and Leading; The Career of Lieutenant General Edward Honor, Fort Eustis, VA: US Army Transportation School, 2003.
[273] 566th group interview.
[274] Wayne Patrick interview by Richard Killblane at New Orleans, LA, 12 July 2002; Wayne Patrick email to Richard Killblane, August 29, 2005. LTG (R) Edward Honor remembered the gap between serials was 10-15 minutes. Edward Honor email to Richard Killblane, August 31, 2005. Wayne Patrick remembered that the gap was 4-6 minutes, Wayne Patrick email to Richard Killblane, September 1, 2005.
[275] Patrick email.
[276] Patrick interview, 12 July 2002 and 17 June 2005 and email August 29, 2005.
[277] Patrick interview, 12 July 2002 and 17 June 2005.
[278] Patrick email, August 29, 2005.
[279] Patrick interview, 12 July 2002 and 17 June 2005.
[280] Patrick interview, 12 July 2002 and 17 June 2005 and email, August 29, 2005.
[281] Patrick email, August 29, 2005.
[282] Patrick email, August 29, 2005.
[283] Ibid.
[284] Ronald Smith interview by Richard Killblane, 17 June 2005; Robert Dalton email to Richard Killblane, September 6, 2005; and 566th group interview.
[285] Jack Buckwalter telephone interview by Richard Killblane, 7 September 2005; and 566th TC interviews.
[286] 566th group interview, Ronald Smith and Buckwalter interviews.
[287] Ronald Smith interview.
[288] Smith interview and Dalton email.
[289] Buckwalter interview.

Robert Fromm next to convoy of the 585th Transportation Company prior to its move to Phu Bai. Photo courtesy of Robert Fromm

3. I Corps Tactical Zone

In December 1967, US Army Vietnam (USARV) sent two truck companies north into I Corps Tactical Zone to support the recent movement of the 1st Cavalry Division into that area. The 57th Transportation Battalion arrived in February 1968 and set up headquarters at Gai Le Combat Base in Phu Bai on 9 February and then Dong Ha Combat Base on 23 February to support the 1st Air Cavalry and 101st Airborne Divisions just ten miles south of the Demilitarized Zone (DMZ). The battalion initially provided command and control over the 57th Light Truck, 61st Medium Truck Petroleum, 446th Medium Truck, and 585th Medium Truck Companies with a platoon each from the 360th and 538th Medium Truck Petroleum Companies. The 63rd Light Truck, 363rd Light Truck, 572nd Medium Truck, and 863rd Light Truck arrived in March. The battalion received seven armored M35A2 2½ - tons, which it issued to the 57th, 446th and 572nd Transportation Companies.[290]

The 2nd Platoon, Headquarters and Maintenance Sections of the 585th Transportation Company had relocated to Da Nang on 14 January 1968 and then further to Gai Le on 19 January where it conducted port and beach clearance at Hue and Tan My. In March, the rest of the Company joined it in Gai Le.[291]

The truck companies in I Corps Tactical Zone would develop their escort procedures differently than 8th Transportation Group in Northern II Corps. Unlike 8th Group, the truck units did not build gun trucks but relied upon tanks and infantry for escort like the US Marines. When the Marine convoys expected enemy contact, they assigned either a Marine company or platoon to escort the convoy. They loaded the Marines in three separate trucks interspersed through the convoy front, middle and rear just like COL Ludy did with his gun trucks. The purpose of the Marines was to dismount and engage the enemy while the convoy cleared the kill zone. If they anticipated the worst, the Marines integrated tanks in the convoy. Wayne Chalker described a convoy of the 585th Transportation Company that also used tanks as escorts.

Vietnam War 117

12 April 1968
585th Transportation Company
"The Fire Base Bastogne Convoy 4/68" [292]
Wayne Chalker

During the week of April 12, 1968 eight members of the 585th Transportation Co. (Medium Truck Cargo) 39th Transportation Battalion, volunteered for a mission to bring much needed ammunition and gun barrels into FB Bastogne from their base camp at Phu Bai. This is the story from three of the eight who took part in this convoy.

On January 14, 1968, Headquarters, Maintenance, and 2nd Platoon of the 585th were loaded onto LST 551 in Qui Nhon harbor and sailed for Da Nang for ultimate relocation to Camp Eagle, 101st AB base camp, Phu Bai. The remainder of the 585th would join the lead elements sometime in March 1968 following the same procedure.

Prior to the 585th's redeployment to Phu Bai in I Corps, the 585th was part of the 27th Trans Battalion running convoys to exotic places like An Khe, Pleiku, Dak To, Bong Son, Phu Cat, etc from their base camp at Phu Tai in II Corps.

The 585th was a 5-ton tractor-trailer unit. Like most transportation companies in Vietnam, we were self sufficient and very mobile. We were a very close company and depended on each other very much. In mid-April 1968, SGT Edwards, our 1st Platoon SGT, asked for volunteers for a convoy. Eight members volunteered to the best of our knowledge. Six of these volunteers were Wayne C Chalker, Marion Amos, Steve Plummer, Neil Maas, Ernie Monaghan, and James McGrath.

SGT Edwards would not tell us where we were going, but told us to be ready the next morning. "Never before had we ever been asked to volunteer."[293] "We were loaded with artillery projectiles and black powder at the Phu Bai ammo dump."[294] At least two of us, Marion Amos and me, were loaded with 175mm howitzer barrels. "The 101st guys seemed surprised that none of us had assistant drivers with us. I stated that our company was always under strength and that we always drove by ourselves. He told me that where you guys are going, you need to have someone riding shotgun with you, especially with black powder because 'Charlie' gets nervous when he sees this ammo coming toward him. I really didn't know what he meant at the time."[295]

We were going to Fire Base Bastogne which was at the mouth of the A Shau Valley. FB Bastogne conducted fire support missions for units operating mostly in and around the A Shau Valley. FB Bastogne was located west of Phu Bai in Thua Thien Province. We were going to be the first convoy from our company to attempt a resupply of FB Bastogne.

"The eight tractor-trailers were led from the ammo dump through Camp Eagle and into the countryside by a 101st AB jeep. The farther we drove, the heavier the jungle became. I thought it was unusual that there were no civilians in the area. We went around a bend in the road and met up with tanks and APC's that were interspersed with our trucks. I started to worry and put on my helmet and flak jacket."[296]

The OIC of the armor advised us that we would have escort for the last mile or so into FB Bastogne. He stated the left side of the road was under 101st control, while the right side was not. The OIC also told us NOT to stop under any circumstances or we would simply be pushed off the road by the armor in back of that truck.

"The road was turning into a steep narrow path with jungle growing up to the edge of this so called road. In II Corp convoys through 'Ambush Alley', we were told to keep lots of space between trucks so to lessen the effects of an ambush. With this in mind, I dropped back from the APC in front of me. A soldier from the armor behind me ran up and said I was going to get all of our asses shot off if I didn't stay close to the other APC. It then struck me that our survival depended on firepower from the armor."[297] This so-called road, which turned into a path through the jungle, had several steep hills before ending at Bastogne. "Because of the steepness on one hill and my overloaded trailer, I was pushed by an APC."[298]

"Suddenly, all hell broke loose."[299] The armor opened up with everything they had on the ride side of the road. "The noise was incredible. I started seeing red flashes flying over the hood of my truck and exploding on the other side of the road. I could also see small arms fire hitting hear me. I panicked and almost jumped from my truck into the jungle for cover. I'm sure this would have ended everything for those and me in the armor and trucks behind. The Army taught us that you always try and drive through an ambush. Somehow, I got down as low as I could behind the wheel and kept driving. Believing that I was about to die, I remember thinking about how bad my parents were going to feel, and I started to pray."[300]

In my truck, I remember the intense firing, but I also remember that I had to get my truck up and over each hill in front of me. Survival, in my mind, was to stay close to the tank in front of me. As mentioned before, I was carrying 175mm howitzer barrels. The weight was far too heavy for my 5-ton tractor. While going up one steep hill, I reached '1st under' in a very short time. Even at 1st under, I felt my Rpm's drop.

As I neared the crest of this hill, even over the firing, I knew something happened to my engine. The thought of being pushed off the road by the APC behind me was frightening. I said a prayer and miraculously my engine held out to the top of the hill.

Not long after, all our trucks made it safety into FB Bastogne. "The guys who unloaded us at Bastogne seemed surprised we arrived intact. One asked me if I had any RPG's fired at me. I said I don't know, what are those? He said they would have been red flashes and explosions after they hit. I said yes, several and I wondered what those were. They also asked which side of the road the incoming fire came from. I said the left. They told me that the left side was supposed to be secure."[301]

Having never been to a firebase like Bastogne before, I got a bad feeling about this place and was very anxious to get unloaded and back on the road again. The jungle came right up to the perimeter of the base. Sort of like an island in the middle of the Pacific.

We were hurrying to get unloaded and regroup for our fun drive back, when the Major (OIC) informed us the road back has been closed. Just prior to this, Steve remembered a small convoy of duce and a half's leaving Bastogne just after we arrived and escorted by the same armor that brought us in. It was rumored that this convoy was hit hard going down the same road we just came up.

The OIC told us to spread out our trucks in the firebase so we would be a smaller target. The Major also advised us to be ready to move out on 15-minute notice. We were to remain at Bastogne for the next six days waiting for the road to be re-secured by the 101st.

We were obviously unprepared for our unscheduled stopover at Bastogne. None of us had any change of clothes, food, personal hygiene items, or extra ammunition. During the week, we experienced what life was really like at a firebase. "Sitting and waiting all those days, I'd see jets dive and drop napalm near the perimeter or on the hillsides."

We caught up on much needed sleep during this week. I remember pulling guard duty with a 101st guy and trading hand grenades with him. I had the newer, baseball type and he had the older oval style. (In addition to our personal issued weapons, 585th members carried hand grenades in their trucks). During the week, the base ammo dump blew. "We jumped into bunkers with the 101st people. The ammo was still going off when I saw at least four medics with stretchers running towards it. I think that was the bravest thing I ever saw. No one must have been hurt because when it was all over, they walked back."[302]

On or about April 24th, the Major told us to pack up and get ready to move out in 15 minutes. Our stay at Bastogne was ending. We were happy to be leaving and getting back to base camp, but were fearful of what lie ahead. "Many of us were worried about the road back, and I remember some of us were saying good bye to our friends just in case."[303]

The return convoy was set up similar to the one we came in with a week earlier, tank, truck, APC, truck, etc. I remember the Major asking us if we could assist his armor with suppression fire on the left side of the road. I thought this odd, as all of us know it is hard to shoot and drive at the same time.

Soon after leaving Bastogne, the firing commenced again. The intensity was the same as coming in the week before. I remember several explosions on the edge of the road to our left. I was doing my best to fire out the window and steer at the same time. I remember shell casings from my M-14 burning my left arm.

"We left the fire base and started driving down the mountain when all hell broke loose. The tanks and APC's opened fire and we started shooting to the left of the road. I had my rifle cradled in my left arm and I was shooting out the driver's side window. It was a real challenge shifting gears and shooting out the window and not running into the back of the APC in front of me. When one magazine emptied, I would put in another and keep firing."[304]

"After checking my forward movement, I looked to the left, and continued to fire my weapon. This is when I thought the devil himself had just hit me between the

Vietnam War 119

eyes with his fist. My head jolted and snapped back. My black plastic rim glasses were shoved back and down into my nose. The pain of being hit between the eyes was excruciating. I thought my nose was broke. I wasn't sure what happened. So many things run through your mind. First, I thought I must have hit a pothole and bumped my head on the steering wheel. When I looked up, everything was black. I looked around and saw nothing but darkness. A few seconds passed and my vision returned. Everything happened so fast that I was in a state of confusion for a moment. After realizing I didn't hit a pothole, I gathered my thoughts, pushed my glasses back up on my nose and kept shooting and driving. The only thing on my mind again was to get out of that area as quick as we could."[305]

Marion's truck was ahead of mine in the convoy back. As I previously mentioned, I remember several explosions just in front of my truck on the left edge of the road. One of these explosions would have been in line with the driver's side of Marion's truck. When we cleared the fire zone area, the armor pulled out and we pulled off and stopped to regroup. Marion came walking back to me and had blood streaming down his face from a hole in his forehead just above the rim of his glasses.

He had obviously been hit with shrapnel and it had been deflected off the center portion of his glasses. This may have saved his live or at least prevented a more serious wound. I knew we had to get out of this area as quickly as possible. We were still in hostile territory without any armor or convoy protection.

I sat Marion down and bandaged him as best I could and asked him if he could drive. He said he could. I told him to stay in front of me so I could watch him until we arrived back at Phu Bai which was 10-15 miles away. In addition to no convoy protection, we had no radio and no OIC.

The convoy into Bastogne the week before began to take its toll on our trucks. "We all knew that this area was unsafe and everyone wanted out of there as fast as possible, but more trucks started breaking down. I think we may have had up to four trucks no longer running by the time we got back to Phu Bai. We came in pushing and pulling each other at about 5 miles per hour. No one was left behind. We all came back together."[306]

When we arrived back at Phu Bai, I escorted Marion to an aid station. The medic examined him, re-bandaged him and told him to go back to work. Several days later, Marion experienced severe headaches and returned to a Marine aid station. There he was re-examined and some shrapnel was removed from his forehead.

Marion, to this day, carries a small piece of shrapnel in his head from that day. Nothing more was ever said to the eight from that convoy. We all went back to work the next day hauling to Camp Evans, Quang Tri, Dong Ha etc. No one, to our knowledge, ever received any award or commendation for our volunteer mission into Bastogne. Marion did not even receive a Purple Heart for his wound.

Lesson
This was one of the rare cases where tanks escorted a convoy. The important factor was that the tanks and APCs remained with the convoy and did not abandon it to chase after the enemy. The steep jungle terrain gave them no other choice. One might also notice there was a tank or APC between each truck. That provided enough fire power to beat back even the most determined attack. This type of escort was only practical because of the high probability of enemy ambush and importance of the cargo. With enough fire power a convoy can travel down any road regardless of the threat.

39th Transportation Battalion

The number of truck companies required a second truck battalion and the 39th Transportation Battalion assumed control of all the truck companies except the 63rd Light Truck in May 1968. The 57th Battalion would support US Army units in the southern part of the I Corps Tactical Zone and the 39th Battalion would support the US Army units in the northern half of the I Corps Tactical Zone. In early 1970, the battalion commander of the 39th Transportation Battalion required the truck companies to construct gun trucks, but they built only one per company probably because the small convoys of 10 to 15 vehicles did not require more gun trucks. Most convoys were escorted by gun jeeps and a gun truck and gun jeep was assigned to the convoy routes where they expected contact. For the most part ambushes were small in comparison to those along QL19 and most were restricted to Hai Van Pass or heading out to the remote fire bases. So one gun truck per company proved sufficient until Operation Lam Son 719 in January 1971.

Ambushes 1971
- Lam Son

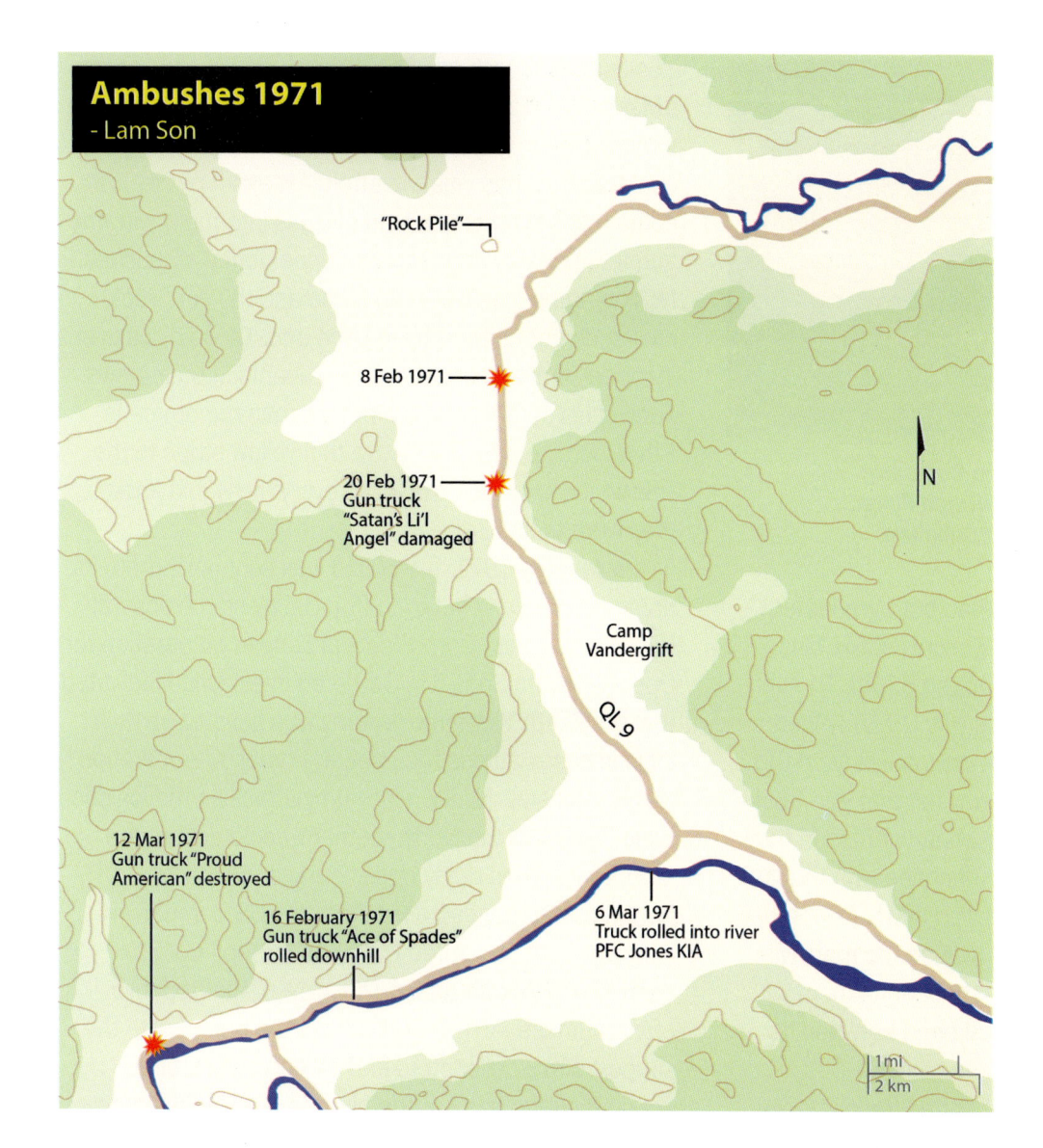

Map annotations:
- "Rock Pile"
- 8 Feb 1971
- 20 Feb 1971 Gun truck "Satan's Li'l Angel" damaged
- Camp Vandergrift
- QL 9
- 12 Mar 1971 Gun truck "Proud American" destroyed
- 16 February 1971 Gun truck "Ace of Spades" rolled downhill
- 6 Mar 1971 Truck rolled into river PFC Jones KIA

to the barracks at night.

This late in the war, there was a shortage of Transportation Corps officers. Second lieutenants made first lieutenant in one year then captain in the next. The rapid promotion and other duties caused a shortage of lieutenants in the truck companies, but the drawdown of Infantry divisions created an excess of infantry officers. So the burden of leading convoys fell heavily on the NCOs. However, during 1970, the Army assigned three infantry officers to the 523rd. Ranger qualified, 1LT Ralph Fuller had recently served in the 25th Infantry Division but when it was inactivated, he still had part of his one-year tour to complete, so he was assigned to the 523rd.

Lam Son 719

By 1970, the gun truck design and doctrine had reached fruition. Experimentation had ended. The 523rd Transportation Company (Medium Truck) had six gun trucks; Satan's Li'l Angel, Ace of Spades, Black Widow, Uncle Meat, King Kong and Eve of Destruction. Each of the three platoons had two gun trucks. By then, new crew members were volunteers selected by consensus of the other crew members. The gun truck crews felt elite but the 523rd believed that by assigning two gun trucks to each platoon, rather than to their own platoon, they felt more like members of the company. They lived with the drivers who they had to protect. Since they were the best, much was expected of them and if they failed to defend the trucks then they would have to face their brothers when they returned

M54 5-ton cargo trucks of the 523d Transportation Company lined up at the unimproved portion of the QL9 heading from Camp Vandergrift to Khe Sanh.
Photo courtesy of Danny Cochran

Vietnam War 121

Officers of the 523d Transportation Company: 1LT Daugherty, 2LT Jim Baird, 1LT Ralph Fuller, CPT Don Voightritter, and 1LT Tom Callahan.

2LTs Jim Baird and Tom Callahan had both graduated from the same Officer Candidate School (OCS) Class 2-70 and were assigned directly to the 523rd. The only logical reason they could conclude why US Army Vietnam assigned Infantry officers to truck companies were the gun trucks. This hinted that the Army felt that gun trucks were a combat arms mission. For whatever reason they were assigned there, the officers identified with the gun trucks and loved the men who crewed them. As combat arms officers they felt their place was on the road. Many preferred to ride in the gun trucks, endearing them with the crews. The 523rd had a special bond between its officers and enlisted.[307]

CPT Donald Voightritter commanded the company. CPT V was a fair and respected officer. His brother Ronald, another Transportation Corps officer, had already earned the Silver Star Medal for valor in the 15 June 1970 ambush. The personality of commanders defines the character of their commands. Voightritter created an atmosphere of mutual respect and camaraderie. This was the strongest asset of the company. The officers would discuss informally with the gun truck crews what they had done during ambushes. No two ambushes were the same and the gun truck crews reacted differently to each one. These discussions inspired confidence with the lieutenants in their crews. Feeling his company was different than the others in 8th Group, he added an extra thin yellow stripe to the yellow noses of their trucks so everyone would know they were the 523rd.

In January 1971, the 523rd was sent north from its home in Cha Rang Valley to Phu Bai along the coastal Highway 1 in the I Corps Tactical Zone in preparation for Operation Lam Son 719. Gathered intelligence indicated that the North Vietnamese Army was building up their logistic bases across the Laotian border in preparation for an offensive. MACV felt that by attacking the logistic base closest to the North Vietnam would disrupt the preparations and delay or discourage the attack. Congress had passed a law after the US Cambodian incursion that prevented US ground troops from crossing the border again. The Army of the Republic of Vietnamese (ARVN) would cross the border with the support of US helicopters and artillery. The artillery set up their fire base at the abandoned Marine camp, Khe Sanh. The 39th Transportation Battalion had the responsibility to supply the forward deployed troops. The east-west, QL9 was the supply route and the battalion

Unimproved portion of QL9. Photo courtesy of the US Army Transportation Museum

The paved portion of QL9 with the Rock Pile in the distance, as seen from the back of King Kong. Photo courtesy of Danny Cochran

122   Convoy Ambush Case Studies - Volume I

posted two 5-ton cargo truck companies, the 515th and 523rd, at the previously abandoned Camp Vandergrift.[308]

To weaken American support to the ARVN drive into Laos, the NVA likewise struck the American supply line with the intent to completely shut it down. Two areas lent themselves to ideal ambush areas, QL9 and Hai Van Pass along Highway 1. During the two and a half month operation, the NVA conducted 23 convoy ambushes along QL9. Three of the ambushes exceeded the others in ferocity and casualties.

The road to Vandergrift was a two-lane paved road through a valley of tall elephant grass up to the Rock Pile, a rocky outcrop that rose 210 meters above the surrounding terrain. From there, a narrow, single lane, unimproved dirt road snaked along the ridge with a river 50 to 100 feet below. The rains could cause the clay road to cave way down the steep embankment below it. The demand for supplies required the two truck companies to deliver cargo around the clock, day and night. To prevent driving off the road at night, the trucks rolled with their lights on giving the enemy ample warning of their arrival. The steep slopes with thick jungle vegetation growing right up to the road made this ideal ambush terrain. With the commencement of Lam Son 719, the guerrillas stepped up the frequency and ferocity of their ambushes hoping to starve off the American support.

8 February 1971
585th Transportation Company

On 3 February, ARVN units finally arrived at Khe Sanh and the next day, American infantry and artillery occupied the abandoned border village of Tabat. On 8 February, ARVN troops would cross the border and invade Laos. Their objective was to advance in a series of air assaults by the 1st ARVN Airborne Division and three ARVN Ranger battalions with ground pushes by the 1st ARVN Armored Division to the Tchepone with QL9 as their supply route. To stall the offensive, the NVA stepped up their attacks on the American convoys along QL9. Bad weather, however, prevented C-130s from flying desperately needed JP4 aviation fuel bladders into Khe Sanh for the start of the offensive.[309]

On 7 February, the 57th Transportation Company (POL) sent a small convoy of tractors hauling empty M49C 5,000-gallon fuel trailers to see if they could negotiate the narrow road to Khe Sanh. The captain arrived enthusiastic that the test worked.[310] Later that day three gun trucks and an M113 escorted a convoy of fifteen to twenty-four 5,000-gallon fuel trailers of the 3rd Platoon of the 57th Transportation Company to Vandergrift. The convoy departed Quang Tri at dusk.

Eve of Destruction of the 523d Transportation Company as it looked during Lam Son 719. Photo courtesty of Fred Teneyck

The Protector belonged to the 57th Transportation Company. Photo courtesy of Jose Garcia and James Lyles

The gun trucks came from three different companies.

The lead truck was a regular cargo truck that knew the route followed by four fuel tankers and then a gun truck and five more tankers. 1LT Roger Maloney remembered the Eve of Destruction from the 523rd Light Truck Company was in the convoy. The Eve had both right and left front mounted .50s and rear twin .50s mounted on a pedestal. It had the unique double steel wall design with filler material in between the walls to reduce the blast effect of an RPG. The entire bed of the gun box was covered with M-2 .50 caliber ammunition. Coming up from QL19 in the Central Highlands, the crew of the Eve had plenty of combat experience. These gunners from 8th Transportation Group used spraying fire instead of short three to six-round bursts. So they would easily run out of ammunition in ten minutes of steady firing.

The Protector belonged to the POL platoon of the 57th Transportation Company and was the second gun truck in line of march. It had two side mounted M-60s and a rear pedestal-mounted M-2 .50 caliber machine gun. SGT Healy's crew of The Protector consisted of Sam Bass manned the .50, Kurtz was the driver. SP4 Charles H. Soule was not a regular crew member, but had wanted to go out on a gun truck so an M60 gunner allowed him to take his place. LT Sam Hoskins was riding as the officer in charge of the gun truck and LT Lee Stafford, of the 572nd "Gypsy Bandits," was just hitching a ride. The next gun truck in line was The Baby Sitters, which belonged to the 515th Light Truck Company.[311]

The 515th only had one gun truck, The Baby Sitters, armed with a forward and rear pedestal-mounted M-2 .50 caliber machine gun and two side-mounted M-60s. SGT Roger Bittner had been the NCOIC of the gun truck since his arrival in Vietnam last June. Bittner's crew had joined about the same time he had, which consisted of Gene Jones as the driver, Walter Crow from Texas was an M-60 gunner, Keith Shed the other M-60 gunner and Pete Aragon had the rear .50. Bittner manned the forward .50. The gun truck also carried a couple M-79s and LAWs, and about 10,000 to 15,000 rounds per machine gun. The gunners either fired short bursts or spraying fire depending upon the situation or how the gunner felt at the time. They had just rebuilt the gun truck in December with a new 5-ton, ballistic glass, single wall of ½-inch armor plating and added the rear .50. Most of their previous enemy contact had occurred in Hai Van Pass but nothing like what they would experience that night. Bittner did not have much use for officers, but hit it off with 1LT Roger Maloney who had joined the company in September. Maloney liked riding in gun trucks and was riding in The Baby Sitters that night.[312]

The Baby Sitters belonged to the 515th Transportation Company. Photo courtesy of Mike Lavin

Maloney remembered The Baby Sitters was up front that night but Bittner remembered it was behind The Protector with ten trucks and two gun trucks ahead of it. There were five tankers between each gun truck with an APC in the rear.

Because the 57th Transportation Company only had three gun trucks, The Protector, Justifier and Assassins, the 1st Brigade, 5th Mechanized Infantry Division loaned it an M113 armored personnel carrier (APC) with a turret-mounted .50 on front and two left and right M-60s in the open hatch in the back. SGT Michael "Mike" McBride, former NCOIC of The Protector, was assigned to it, and picked Leroy Sherrill and another kid named, Spears, as his gunners. Since the convoys ran at about 25 mph due to the condition of QL9, the APC had no problem keeping up.[313]

With gun trucks from three different companies, each had different policies for reaction to contact. The 515th policy was for every truck to clear the kill zone; even the vehicles not in the kill zone were going to drive through it. The 523rd policy was for only the vehicles in the kill zone to clear it, and the gun trucks would remain in the kill zone to protect any drivers in the kill zone. That night the convoy halted by the Rock Pile to reestablish their interval. Not too far away parked a couple M113 scouts from the 1st Battalion, 77th Armor of the 5th Mechanized Infantry Division. Suddenly the convoy received a couple mortar rounds and Maloney saw the muzzle flash from the back side of the hill, so he moved the convoy further down the road. The Rock Pile was on the right side of the road and there was a bend in the road that took a sharp turn to the left. The convoy continued down into a draw, crossed the bridge and staged up again near a short bridge with embankments on both sides. As the convoy slowed down to negotiate the sharp bend in the road at 0115 hours, it was then hit by intense small arms fire from tall grass mostly on the right (passenger) side.[314]

The enemy hit the front of the column as if they knew what the convoy was hauling. The enemy hit a couple tankers from the right (passenger) side of the road and they exploded. The kill zone contained the vehicles between the lead truck and The Baby Sitters. In the dark the gunners could see the muzzle flash and green tracers. Maloney fired M-79 grenades as fast as he could while everyone else blasted away. So Bittner swung his .50 around while firing and shot out the antenna causing him to lose radio communication with everyone else.[315]

An RPG slammed into driver's side of The Protector killing Soule, blew the rest of the crew out and blinded LT Stafford in back. The enemy poured small arms, rockets and mortars into three different parts of the convoy setting three tankers on fire. Bittner saw a lieutenant in The Protector stand up when the second or third tanker was hit; he then fell down as if he had melted. Bittner did not like lieutenants on gun trucks, because he did not trust them to keep their heads down.[316]

The enemy destroyed The Protector and three fuel tankers behind it had exploded. The Baby Sitters, however, cleared the kill zone with 10-12 holes and a couple tires flat on the right side. The one-way road was so narrow there was no place anyone could drive past the tankers except a place next to the bridge wide

QL9 leading west to Rock Pile. Photos courtesy of Mike Lavin

QL9 heading south from the Rock Pile where 8 February ambush took place.

Vietnam War 125

enough for a truck to park. McBride's APC in the rear was undamaged. The sound of explosions startled the 1/77 scouts at the security check point by the Rock Pile. The scouts looked to their west and saw a ball of fire illuminate the area followed by the sound of small arms fire. They cranked up their APCs and saw a lone APC speeding away from the kill zone. McBride said he was leading trucks out of the kill zone. LT Revelak ordered his APC into the kill zone where they saw three burning fuel tankers, one which had run over the embankment. SP4 Harold Spurgeon saw figures moving in the brush but could not distinguish friend from foe so he held his fire. The tracks advanced further into the kill zone and Spurgeon looked down and saw a soldier wearing a helmet running alongside yelling at him, but could not hear what he was saying. He later learned the guy was warning them not to enter the kill zone. Revelak yelled at his scouts to open fire and they sprayed the sides of the road. They drove past burning trucks and across a bridge then saw a burning gun truck ahead of them. Spurgeon saw what he thought was a baldheaded NVA soldier crawling in a ditch ten feet from his APC and was about to shoot him with his M-60 when he realized it was an African-American soldier.[317]

As the APC rolled further into the kill zone, another soldier emerged from the shadows and ran at a crouch beside his APC yelling something up at the crew, which they could not hear. When the track pulled up next to the gun truck, its machineguns hung down and the gun truck appeared empty and lifeless. They saw nothing past this point, so the APC pivot steered in the road and headed back. The lieutenant yelled at his crew to open fire again. They then stopped and dismounted in the kill zone next to one burnt out truck and four others riddled with bullet holes. All the drivers were missing.[318]

The Baby Sitters sat at the bridge letting the other trucks pass. Bittner remembered guilt took over so he wanted to go back in because it was the right thing to do. Maloney remembered telling Bittner and the driver, Gene Jones, they had to go back and help the others because the trucks were getting hammered. They then had to figure out how to turn around on a narrow road in the dark and get back in the kill zone. By the time The Baby Sitters returned the enemy had broken contact. The Baby Sitters found five or six wounded drivers in a ditch beside the road and could hear enemy voices out in the bush.[319]

Meanwhile, the lead element of the convoy

Photo courtesy of Mike Lavin

Award ceremony for 8 February 1971 ambush. Photo courtesy of Mike Lavin

continued to Vandergrift and the medevac came in to take away the wounded. It seemed like it took forever for the medevac to come in because the area still received sporadic small arms fire. After they loaded the wounded in the helicopter, the rest of the convoy continued on to Vandergrift.[320]

The Baby Sitters had to figure out how to get a burning tanker off the road because the road was so narrow; there was no room to pass for drivers to jump on passing vehicles. The rest of the trucks picked up drivers from five damaged trucks and drove in reverse back to Quang Tri.[321]

Maloney remembered CPT Ken Cloucci, of the 585th Transportation Company at Vandergrift, came to Maloney and said he had one driver missing – unaccounted for. So Maloney and CPT Colucci took The Baby Sitters back into the kill zone. There were at least two burning tankers maybe three still in the kill zone. Maloney jumped up on the running board of one and swears he saw the driver in there. However he was not. The injured driver had escaped the vehicle and another truck drove him to the medevac station at Vandergrift. The convoy delayed at Vandergrift for a couple days then continued on to Khe Sanh delivering its load a vital aviation fuel. BG Arthur H. Sweeney, Jr., Commander of US Army Support Command, Da Nang was waiting for them at the gate.[322]

The convoy had nine wounded and Charles H. Soule killed. One gun truck was destroyed, and three tankers and two other gun trucks were damaged. The Protector was repaired and renamed the Executioner. The 57th POL never tried to go back to Khe Sanh because the road did not allow for vehicles to pass in the event of an ambush. From then on the 57th POL ran day convoys from Quang Tri to Da Nang.[323]

Lesson

This ambush hinted at the increasing danger that awaited the truck drivers for the rest of the operation. The elephant grass provided the enemy ample concealment so they could approach right up to the side of the road. As the convoys of the 8th Transportation Group had learned in An Khe Pass, areas flanked by tall elephant grass where convoys had to slow down made good kill zones. Deliberately stopping in an area that provided similar concealment for the enemy invited trouble. While the interval at the time of the halt was not known, the initial kill zone included four vehicles. Because the 515th had never faced an ambush on this scale before the policy of clearing the kill zone left disabled trucks and their drivers undefended in the kill zone for a short period. Fortunately, the APC of a nearby check point arrived in short order to defend the truck drivers of the damaged vehicles. The crew of The Baby Sitters made the right decision to turn back into the kill zone but passing trucks escaping the kill zone on that narrow road delayed its return. This same stretch of dirt road between the Rock Pile and Vandergrift, however, would be the scene of another costly convoy ambush. The next time the enemy would attack a convoy of the 523rd, a hardened veteran of QL19 ambushes.

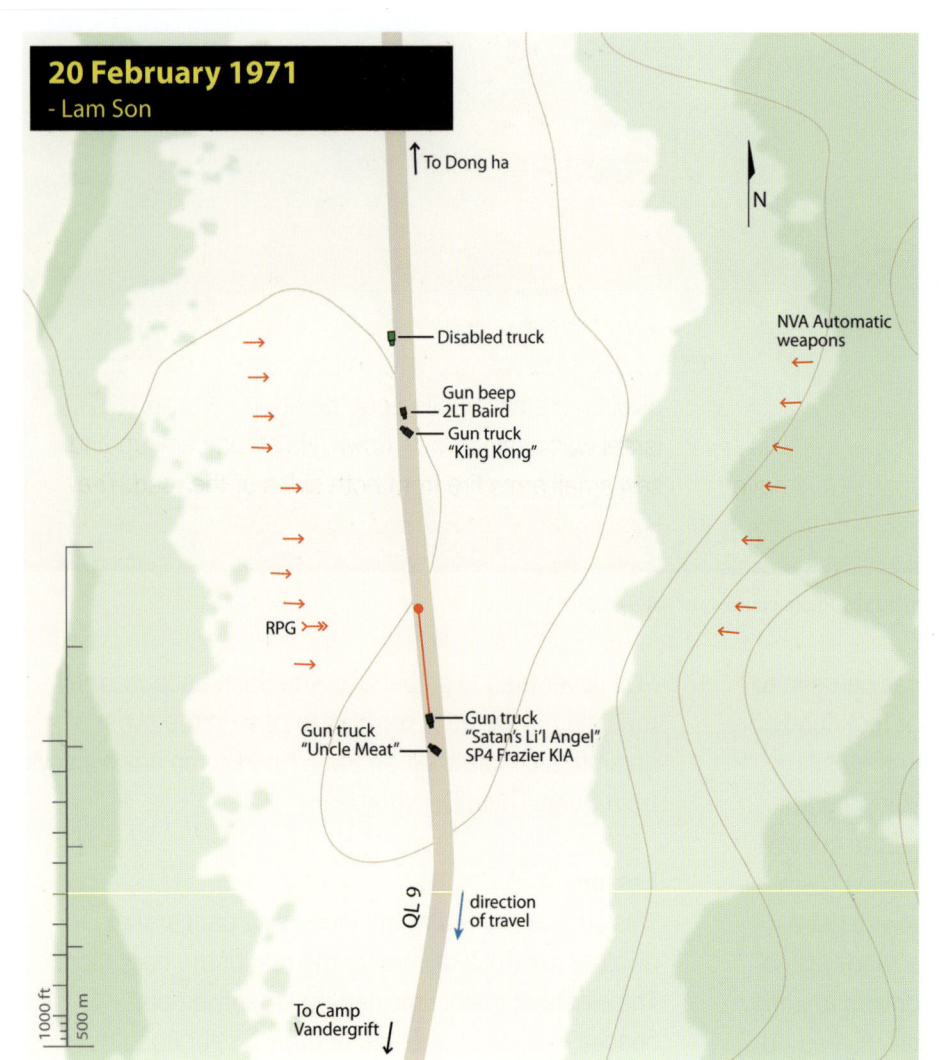

20 February 1971
- Lam Son

Map labels:
- To Dong ha
- Disabled truck
- Gun beep 2LT Baird
- Gun truck "King Kong"
- NVA Automatic weapons
- RPG
- Gun truck "Uncle Meat"
- Gun truck "Satan's Li'l Angel" SP4 Frazier KIA
- QL 9
- direction of travel
- To Camp Vandergrift

20 February 1971
523rd Transportation Company, 39th Transportation Battalion

2LT Baird had been sent back to Phu Bai to pick up 17 brand new 5-ton trucks. They returned after dark. The convoy doctrine at the time was to limit convoys to no more than 30 trucks with a gun truck ratio of 1:10. Uncle Meat and King Kong had escorted his convoy and both gun trucks had three M-2 .50 caliber machine guns. The M-2 .50 was the most successful design in American weapons and had seen very little change in its design since its original issue in 1919. This time Baird rode close to the rear in a ¾-ton gun "beep"[324] with twin M-60 machine guns. He noticed that some Transportation Corps officers preferred to ride up front. He knew that if there was trouble it would invariably occur in

Uncle Meat. Photo courtesy of US Transportation Museum

128 Convoy Ambush Case Studies - Volume I

King Kong. Photo courtesy of Danny Cochran

the rear and that is where the key decisions would have to be made. If an ambush split the convoy, by doctrine the trucks out of the kill zone would continue to roll on to the next security check point or camp. If the convoy commander was in the lead then he would be unable to make the key decisions for the rest of the convoy either trapped in the kill zone or behind. 1LT David R. Wilson was killed trying to reenter the kill zone in an unprotected jeep.[325]

Satan's Li'l Angel had escorted a convoy that had delivered ARVN soldiers to Laos. The crew had been on the road for 36 hours when they pulled into Dong Ha that night. They planned to RON because their convoy could not unload until morning, then 2LT Baird called them on the radio, "Hey Red, I'm a little light on security and expect some movement." They had received a lot of reports of enemy movement along the route and SGT Chester Israel had seen movement on the way down. Baird only had two gun trucks and his ¾-ton at Dong Ha. He needed some "heavy hitters" in the rear and asked Chester if he would go back with him. Chester asked his crew and they volunteered to go back to Vandergrift.[326]

Uncle Meat led with King Kong in the middle, with Satan's Li'l Angel followed by Baird's ¾-ton in the rear. It was dark on 20 February as the convoy neared Camp Vandergrift. The mountain ridge to the east came within yards of Highway 9 and a valley of tall elephant grass covered the valley to the ridge line to the west. Around midnight a mile and a half from their destination, Baird heard an explosion followed by an intense volume of small arms fire from the jungle on the ridge to his left.[327]

An RPG fired from the right (west) side of the road struck Satan's Li'l Angel's gun box right under SP4 Richard B. Frazier's gun killing him instantly. Shrapnel and concussion of the blast knocked Israel to the side of the wall wounding him. The initial blast was followed by small arms fire from both sides of the road, but Satan did not stop. Israel climbed over and checked Frazier but already knew he was dead. He then rose up and looked out at the tracer fire. Calvin Bennett was firing to the right side.[328]

A second RPG came through the right front wheel, slammed into the engine and took out a piston. The engine locked up and the truck lurched into a nose dive, then smoke boiled out the hood and cab. SGT Israel was again thrown down. He stood back up and saw small arms fire from both sides of the road. They were right in the middle of the kill zone. Bennett fired his left rear .50 until it jammed. Frazier's .50 was hit by shrapnel and did not work. Israel's .50 would only fire single shots. Small arms fire shot out the tires of the gun truck. The NVA had learned to take out the gun trucks first before they went after the rest of the trucks so they concentrated their fire on Satan.[329]

Baird's ¾-ton raced ahead and passed a disabled

Satan's Li'l Angel next to Black Widow. Photo courtesy of US Transportation Museum

Baird's ¾-ton. Photo courtesy of Danny Cochran

Vietnam War 129

Where RPG hit gun box on Satan's Li'l Angel. Photos courtesy of Logan Werth

5-ton cargo truck in the ditch. He ordered his driver to stop so they could check on the driver. They came to a halt a hundred feet ahead of the truck. He did not want to leave the disabled truck until he was sure that its driver was safe. To do so required him to wait in the middle of the kill zone. As soon as his gunner tried to return fire, both M-60s failed to fire. Evidently, he had put the gas plugs in backwards when he reassembled them. The three men only had one M-79 grenade launcher and their M-16s to defend against an NVA company. Baird immediately radioed the two lead gun trucks and told them to come back. The one thing that Baird could depend on was the loyalty of his gun trucks to rescue him or any other truck in trouble.[330]

SGT Israel looked back at Baird's ¾-ton and saw it had stopped and was taking fire. It was sitting cockeyed with the wheel in the road. Chester could not see any fire coming from Baird's vehicle. SP4 Robert W. Thorne climbed into the gun box from the cab and asked, "Chester, are you hit?" Israel answered, "I'm fine. The only hope we got is for you to get this started and get us out of here!" Thorne climbed back into the driver's seat and started up the engine. It clanged and another RPG hit the right front rear duals and the truck bounced. Thorne kept that truck going and it crawled about 25 yards, enough to clear the kill zone, before the engine shut down. They continued to return fire.[331]

Neither the crew of Uncle Meat nor King Kong had heard the gun fire behind them. The majority of the convoy had continued to Vandergrift as if nothing had happened. Uncle Meat had already entered the compound and King Kong had just made the right hand turn into Vandergrift when they heard Baird's call for help. They immediately backed up, turned around and raced as fast as their trucks would let them back to the kill zone.[332]

Baird knew his gun truck crews and had confidence in their judgment. He also knew that too much jabber on the radio would cause confusion and tie up the radio net. He quickly and precisely informed the gun trucks of the situation. Satan's Li'l Angel had been hit, his gun

Damage to Baird's ¾-ton. Photo courtesy of Danny Cochran

130 Convoy Ambush Case Studies - Volume I

beep and one 5-ton were still in the kill zone. The crews asked which side of the road the enemy was on and Baird informed them that he was taking small arms fire from the ridge to his left and the field of elephant grass to his right. The enemy was close enough to throw hand grenades at his vehicle. He then quit talking. He would count on their judgment as what to do.[333]

Ten minutes of steady small arms fire had elapsed since the beginning of the ambush. By then Baird was taking fire from both sides of the road. The enemy was closing in from the elephant grass while others fired down on them from the ridge to the east. His gunner, Downer, tapped him on the shoulder and said, "I see one. What do I do?" Baird turned, looked back down the road and saw an enemy soldier about 15 meters away on a berm alongside the road loading an RPG. He told his gunner to shoot him. The gunner fired his M-79 grenade launcher at him. The enemy soldier was too close for the 40mm grenade to arm in flight. It struck him with enough velocity to either kill or incapacitate him, because he did not fire his rocket.[334]

Around ten minutes after the initiation of the ambush, King Kong raced past Satan and up to their convoy commander's ¾-ton, parking right in front of it at an angle facing to the west. Uncle Meat similarly parked near Satan's Li'l Angel. Baird was never as glad as when he saw the tracers of those .50s. There was a reassurance that everything would turn out alright. Baird knew his gun truck crews knew what to do and called on the radio, "They're in the ditches. They're in the ditches."

Where the second RPG hit the front of Satans Li'l Angel.
Photo courtesy of Danny Cochran

The gunners on the Kong swung their .50s around and sprayed the ditches. Kong and Uncle Meat took the pressure off of Satan as the enemy concentrated their fire on the two new gun trucks.[335]

The success of an ambush depended upon surprise and extreme violence. The gun truck crews had learned to turn the fight back on the enemy as fast as they could with even more violence. This would take the psychological advantage away from the enemy forcing them to break contact. The .50s blazed away in four to six round bursts at the muzzle flashes to their left and right. The gunners poured 30-weight oil from plastic canteens to help cool the barrels and ensure the smooth function of their breaches after firing off about three to four boxes of ammunition.[336]

An RPG hit the left rear duals and exploded in all the colors of the rainbow under left rear gunner, Danny Cochran, knocking him backwards on Larson manning the right .50. Cochran then jumped back up, grabbed his .50 and went back to work. King Kong was an APC gun truck. Large chunks of hot shrapnel had come up through the aluminum floor of the hull and lodged in the top of the box right under his machine gun. One piece of shrapnel had burnt a hole in the charging handle and others had left five or six holes in the barrel, but it still fired.[337]

The barrels turned red and as soon as the gunners saw the rounds curve after they left the barrel they knew it was time to change them. Each time Emery, manning the turret gun, swung his barrel toward Cochran, Cochran grabbed the asbestos glove, spun the barrel off then picked up a new barrel and spun it on tight, by feel counted three clicks back and let go. Gunners had different methods of setting the headspace and timing and none used the timing gauge during an ambush. They knew their guns. Some memorized the number of clicks for each breach. Others wrote the number of clicks needed on each barrel. Cochran did it by feel. He backed off the three clicks and depending upon the rate of fire of his .50 he added more clicks. He knew his .50. He changed three barrels for the truck commander and a similar number for himself that night.[338]

The one advantage to fighting at night, the gunners fired in the direction of the enemy muzzle flashes, which betrayed their positions. There was no concealment in the dark once one fired his weapon.

Oil down the side of King Kong. Photo courtesy of Danny Cochran

The tactic worked. After about ten minutes of firing, Uncle Meat and King Kong had turned the fight back on the enemy and they broke contact. During the fight, the driver of the disabled 5-ton had run to his convoy commander's vehicle.

The medic from Baird's ¾-ton ran up and stopped in front of Satan. Bennett remembered the dustoff hovered over the road ahead of their gun truck and threw down a stretcher out one end. SGT Israel and the medic lifted Frazier's body over the cab onto the hood. Blood was everywhere in the truck and they were slipping and stumbling, but got Frazier's body on the stretcher. Israel had several wounds and did not want to leave when the dustoff helicopter landed. The medic hit Chester with a shot of morphine as soon as he got on the ground, and then put him on the dustoff.[339]

That close to Vandergrift, Uncle Meat loaded the rest of the crew from Satan's Li'l Angel into their gun truck then drove off the road and backed up to Baird's vehicle. The drive shaft had broken and the vehicle could not drive. The crew of Uncle Meat hooked up the ¾-ton to Uncle Meat, which towed it into Vandergrift. The two gun trucks that came to the rescue also received damage but could roll under their own power. After the initial volley of fire, no other casualties were taken. King Kong limped back to Vandergrift on its rims.

The sweep of the area the next day discovered four enemy dead and one wounded NVA soldier 25 meters from the road. The enemy usually made great effort to recover their dead and conceal their losses, so no one could accurately determine the total enemy losses. These were the only confirmed enemy kills by the 39th Transportation Battalion Soldiers during Lam Son 719.

Lesson

The success of the ambush depended upon a number of factors. First and foremost, the dedication and determination of the gun truck crews. Without hesitation the two lead gun trucks raced back into the kill zone. The convoy commanders knew their men and trusted in their judgment to do the right thing in an ambush. This allowed Baird to keep his radio transmission short and to the point, keeping the net free for more important traffic. Too much chatter would prevent someone from interjecting important information at the critical time. Each of the gun trucks instinctively raced to a disabled vehicle to provide covering fire. Failure to test fire the M-60s, however, failed to warn the gunner of the ¾-ton that he had reassembled them wrong. On the other hand, the intimate knowledge of the gunners with their .50s caused them to fire nonstop for ten minutes. The .50 caliber round could penetrate through most

Award ceremony for ambush 20 February 1971. Photo courtesy of Logan Werth

trees and bunkers and cut a man in half if it struck him. The hail of six .50s inspired fear in the NVA and caused the enemy to break contact.

Since there was a rumor the enemy had a bounty on gun trucks, the company mechanics repaired Satan's Li'l Angel to get it back out on the road as soon as they could. Bob Thorne and Calvin Bennett remained on the gun truck and the new crew felt the name Satan's Li'l Angel was jinxed, so they repainted it and renamed it Proud American, but it was hit in an ambush before they had time to paint the name on it.

132 Convoy Ambush Case Studies - Volume I

12 March 1971
39th Transportation Battalion

Because the road past Vandergrift could only support 2½ and 5-ton cargo trucks, the 39th Battalion would need another light truck company to push from Vandergrift to Khe Sanh. The thick jungle that grew right up to the road made it ideal for ambushes and the fact that convoys had to run both day and night made ambushes easier. The 1st Brigade, 5th Mechanized Infantry Division had responsibility for the security of that section of the road.

On 12 March, 2LT Tom Callahan, 1st Platoon Leader of 523rd Transportation Company had returned on a convoy from Khe Sanh and informed LTC Alvin C. "Big Al" Ellis, Commander of the 39th Transportation Battalion, his convoy had received enemy fire and thought that it was not safe to take another convoy up there especially since it was going to be dark by the time they returned to Khe Sanh. Regardless, the convoy kicked out from Vandergrift to Khe Sanh with 2LT Jim Baird, 3rd Platoon Leader, as the convoy commander. Just in case the enemy tried to ambush a convoy, the detail left behind kept a reaction force. 1LT Ralph Fuller had all the remaining gun trucks lined up ready to go while he monitored the radio for contact.[340]

2LT Baird led a convoy from Vandergrift to Khe Sanh. On the Proud American, Robert Thorne was the driver, Calvin Bennett was the rear gunner, McDonald was the right gunner, and Nelson Allen was the NCOIC. Both CPT "V" and LT Baird rode on the gun truck. A B-40 rocket hit the gun truck, Proud America, between the cab and the gun box on the driver's side mortally wounding the driver, SP4 Thorne. Thorne steered Proud American into the hillside instead of down the steep cliff into the creek. This saved the rest of the crew. Unfortunately, Baird had been kneeling by the radio mounted in the left front corner of the box when the rocket hit and received multiple fragmentary wounds to his left arm. McDonald, the right front gunner, received a facial wound. Two tankers were immobilized.[341]

Fuller heard the call, "Contact, contact, contact," on the radio and led his convoy of gun trucks. Riding in Daughter of Darkness, an engineer stopped Fuller, "Don't go up there! They're having an ambush! They're having an ambush!" To which Fuller replied, "That's what we're going up there for." About six gun trucks rolled in

Damage to Proud American. Photos courtesy of Logan Werth

Photo courtesy of Danny Cochran

Photo courtesy of US Transportation Museum

and laid down suppressive fire which quickly ended the enemy fire. Fuller saw that Baird was badly wounded, and Baird said Thorne was dead. Callahan laid Baird on a stretcher and Fuller called for a medevac then looked around for a good place for the dustoff helicopter to land. He found an opening by a bridge and Uncle Meat drove Baird to the bridge. The helicopter arrived but was afraid to land; instead, the men lifted the stretcher up to the bird. Fuller told the medevac crew, "Take care of him. He was a good one." They placed Thorne's body in the Black Widow and took it back to Vandergrift. A squad from the 4th Battalion, 3rd Infantry went out to search for the enemy but found no enemy killed or wounded.[342]

Lesson
Anticipating trouble, the 523rd Transportation Company assembled all its gun trucks as a reaction force rather than wait upon the infantry to arrive. Again and again, quick reaction and overwhelming fire power seemed to bring an end to any ambush or buy enough time for the infantry and armor to arrive and take over the fight.

Calvin Bennett had survived two bad ambushes and lost three crew members. Normally one devastating loss like this would justify getting off a gun truck and

Proud American was renamed Ace of Spades in memory of the Ace of Spades that rolled down into the river bed.
Photo courtesy of Fred Teneyck

pulling camp duties until the end of one's tour, but Calvin Bennett remained on his gun truck. Trusting in his instincts, he just did not feel it was his time to die. The new crew, however, decided to name their gun truck Ace of Spades in respect for the Ace of Spades that had been destroyed when the road gave way and it rolled down into the river bottom.

134 Convoy Ambush Case Studies - Volume I

[290] LTC Chester T. Henderson, Operational Report of the 57th Transportation Battalion (Truck) for Quarterly period Ending 30 April 1968, 30 April 1968. Report Control Symbol CSFOR-65 (R-1).
[291] 1LT Thomas L. Tish, Annual Historical Summary, 585th Transportation Company (Medium Truck Cargo), 1 January 1968 – 31 December 1968.
[292] Wayne Chalker, "The Fire Base Bastogne Convoy 4/68," Army Transportation Association Vietnam, 134.198.33.115/atav/default.html.
[293] Steve Plummer, personal account.
[294] Ibid.
[295] Ibid.
[296] Ibid.
[297] Ibid.
[298] Ibid.
[299] Ibid.
[300] Ibid.
[301] Ibid.
[302] Ibid.
[303] Ibid.
[304] Marion Amos, personal account.
[305] Ibid.
[306] Ibid.
[307] Jeffrey Fuller interview by Richard Killblane at Ft Eustis, VA, June 2004, Jim Baird interview by Richard Killblane at Branson, MO, 20 June 2005.
[308] LTC Alvin C. Ellis, "Operational Report – Lessons Learned, 39th Transportation Battalion (Truck), Period Ending 30 April 1971, RCS CS FOR-65 (R2)," Headquarters, 39th Transportation Battalion (Truck), APO San Francisco 96308, 4 May 1971.
[309] Robert Cody Phillips, Operation Lamson 719; The Laotian Incursion During the War in Vietnam, Masters Thesis, May 1979.
[310] Keith William Nolan, Into Laos; The Story of Dewey Canyon II/ Lam Son 719; Vietnam 1971, Novato, CA: Presidio Press, 1986.
[311] The Protector, 57th Transportation Co Nam Nomads, http://nam-nomads.tripod.com/TheProtector/protector.html; and Claude Roberson telephone interview by Richard Killblane, 22 July 2013.
[312] Roger Maloney interview by Richard Killblane at Kirkland. MO, 28 June 2013; and Roger Bittner interview by Richard Killblane at Kirkland. MO, 28 June 2013.
[313] Michael McBride interview by Richard Killblane, 9 July 2004.
[314] Maloney interview, and Roberson interview.
[315] Bittner interview, and Maloney interview.
[316] Bittner interview; and Bill Hampton, "Gypsy Bandits Move to Quang Tri," ATAV, http://grambo.us/atav/hampton.htm.
[317] Nolan, Into Laos.
[318] Nolan, Into Laos.
[319] Bittner and Maloney interview.
[320] Maloney interview.
[321] Bittner interview; and McBride interview;
[322] Maloney interview.
[323] The 5th ID(M) After Action Report stated one vehicle destroyed, five damaged, one driver killed and nine wounded.
[324] The slang term for a ¾-ton gun truck was gun beep because it was bigger than a jeep.
[325] Jim Baird interview, and Danny Cochran interview by Richard Killblane at Branson, MO, 18 June 2005.
[326] Chester Israel and Calvin Bennett interview by Richard Killblane, at Pigeon Forge, TN, 8 August 2008.
[327] The 504th MP Bn Forward Log recorded the ambush at 0015 hours on 21 February 1971, but Baird and Israel remembered it at 2355 or midnight.
[328] Israel and Bennett interview.
[329] Israel and Bennett interview.
[330] Baird interview.
[331] Israel and Bennett interview.
[332] Baird and Cochran interviews.
[333] Ibid.
[334] Ibid.
[335] Ibid.
[336] Cochran interview.
[337] Ibid.
[338] Ibid.
[339] Israel and Bennett interview.
[340] Fuller interview, and Baird interview.
[341] Fuller interview, Bennett interview, and Baird interview.
[342] Fuller interview.

Convoy of the 7th Transportation Battalion. Photo courtesy of US Army Transportation Museum

4. III Corps Tactical Zone

25 August 1968
7th Transportation Battalion, 48th Transportation Group

The 48th Transportation Group consisted of two battalions stationed at Long Binh in III Corps Tactical Zone. The 6th Battalion had all light truck companies and the 7th Battalion had all the medium truck companies. Long Binh cleared cargo from Saigon and the military terminal at Newport and then delivered to base camps throughout the region. This southern part of the country gently sloped downward toward the Saigon and Mekong Rivers. Consequently, the unimproved roads that spread out from Long Binh and Saigon like the spokes of a wheel were flat and filled with pot holes.

Late summer was the monsoon season, so low cloud ceiling with intermittent rain impeded air support the day of the ambush. On 25 August 1968, 81 trucks from the 48th Transportation Group departed Long Binh in three serials with six refrigeration (reefer) trucks in the front, followed by general cargo trucks, then fuel and ammunition in the rear. 11 of the trucks belonged to the 6th Battalion and the rest to the 7th. If the enemy disabled a fuel tanker or ammunition trailer, the first half of the convoy could still escape. C Company, 720th MP Battalion provided six and the 25th MP Company provided two gun jeeps with M-60 machine guns as escorts. The truck drivers carried a basic load of 100 rounds for their M-16s. This convoy was destined to resupply the 1st Brigade, 25th Infantry Division at Tay Ninh 45 miles north of Saigon and the short trip normally took six hours to complete because of the mandated convoy speed limit of 20 miles per hour.[343]

The convoy proceeded west along QL1 from Saigon past the 25th Infantry Division base camp at Cu Chi. There, the convoy divided into two serials and proceeded on to Go Dau Ha at the intersection of QL1 and QL22. The convoy then turned northwest on QL22 to the village of Ap Nhi just 20 miles short of its destination at Tay Ninh.[344]

SP4 David M. Sellman, of the 62nd Transportation Company, was delayed at Cu Chi due to a flat tire and his convoy left without him. After changing the tire, he convinced his superiors he could catch up and raced to join his convoy just before it reached Ap Nhi. As the convoy approached within a mile of the village, the drivers saw many villagers heading south with their possessions and many shouted, "VC." Since the convoy had never faced a large ambush, it took no additional precautions.[345]

The 1st and 3rd Brigades of the 25th Infantry Division normally provided security along the main supply route, but the new division commander, MG Ellis W. Williamson, had ordered the 3rd Brigade back to Saigon. The reduction in force resulted from the anticipated third phase of the Tet Offensive. From 17 to 24 August, the 1st Brigade had fought off 13 enemy battalion or regimental attacks, including seven attacks on the 1st Brigade's bases. The 1st Brigade's Intelligence Officer had determined that 16,000 combat-ready troops of the 5th and 9th NVA Divisions accompanied by an anti-aircraft battalion and two attached VC battalions would mass against it. The division's intelligence officer, on the other hand, believed Saigon was the main target. MG Williamson then moved the 2nd Battalion, 34th Armor back to Cu Chi, while still ordering the 1st Brigade to secure the main supply route. This proved a fatal mistake. The brigade commander, COL Duquesne "Duke" Wolf, doubted whether he could defend his six bases, let alone the main supply route with his meager force. His brigade only had three understrength rifle companies, three understrength mechanized infantry companies, two 105mm artillery batteries, and two medium batteries of 155mm with no armored cavalry units attached. Wolf challenged Williamson's decision, to no avail. Although convoy security was a high priority, MACV respected the commander's decision. So just C Company, 4th Battalion, 23rd Infantry provided security along the route around Ap Nhi and the Army of the Republic of Vietnam (ARVN) had an outpost also at the entrance to the village.[346]

The village of Ap Nhi and the Ben Cui Rubber Plantation, known locally as Little Rubber, flanked QL22 for about a mile. The village of Ap Nhi on the southwest side was mostly farm land while the rubber trees on the northeast side grew to within 15 feet of the road with a drainage ditch and an earthen berm paralleling the road inside the tree line. About four companies of the 88th NVA Regiment had moved into the Little Rubber and another into the village the evening prior to the ambush. About 1145 hours the convoy entered the quiet village of Ap Nhi. It was misting and raining and the ceiling hung about 200 feet above the ground. The convoy passed what looked like a group of Army of the Republic of Vietnam (ARVN) soldiers in camouflage jungle fatigues and jungle boots armed with M-16 rifles along the north side of the road adjacent the Little Rubber firing into the village. The lead vehicles of the convoy started to leave the village and the ammo and fuel vehicles were alongside the column when the supposed ARVN soldiers opened fire on the convoy. They turned out to be Viet Cong.[347]

The enemy hidden in the rubber plantation then initiated the ambush with an intense barrage of rocket, mortars, and automatic small arms fire on the convoy. The enemy first targeted the eight gun jeeps and then fired at the lead fuel trucks setting two on fire hoping to block that part of the convoy. 30 trucks in front of them sped away, following policy, leaving 51 trucks stranded in the mile-long kill zone. The enemy then set two ammunition trailers hauling 105mm rounds on fire at the rear of the convoy sealing the trucks in place. The drivers climbed out of their vehicles and took up defensive positions either behind their

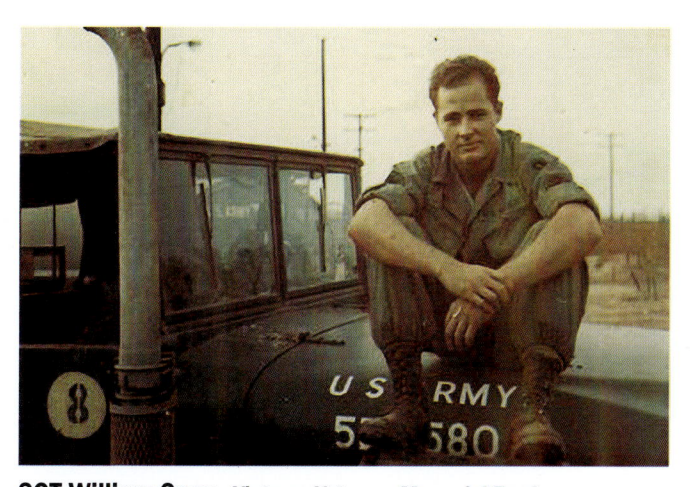

SGT William Seay. Vietnam Veterans Memorial Fund

trucks or in the ditch along the southwest side of the road. In their haste to exit their vehicles, some found themselves without weapons or radios. Others had no water or ammunition since they neglected to wear their pistol belts. The enemy had thoroughly planned the ambush and it occurred well beyond the range of the 1st Brigade's artillery. Likewise, the low ceiling initially prevented the use of air support. With the convoy trapped, the enemy left their cover and made a rush on the column of trucks.[348]

MPs, SSG Charles H. Frazier and SP4 Albert Murphy, dismounted their gun jeep in the first march serial and picking up another wounded MP, dragged him toward a small hut just off the road. Enemy then rushed them from front and rear but the two MPs turned back the assault with their M-16s and defended their position for 30 minutes despite their wounds received until MPs from the 25th MP Company reinforced them.[349]

When the convoy stopped, SP4 William W. Seay, from the 62nd Transportation Company, immediately dismounted and took cover behind the left rear dual wheels of his truck. Seay's trailer carried high-explosive artillery powder charges. SP4 Sellman, in the truck behind Seay, did likewise. Another driver joined him and the three fought about 20 feet apart. When the initial assault reached within ten meters of the road, Seay, who was the closest, opened fire, killing two of the enemy. Sellman shot one just 15 meters in front of him, then his M-16 jammed. Fortunately, the three drivers had successfully turned back the first enemy assault.[350]

The beleaguered drivers then came under automatic fire from the berm and sniper fire from the trees. Seay spotted a sniper in a tree approximately 75 meters to his right front and killed him. Within minutes an enemy grenade rolled under the trailer within a few feet of Sellman. Without hesitation, Seay ran from his covered position while under intense enemy fire, picked up the grenade and threw it back to the enemy position. Four enemy soldiers jumped up from their covered position and tried to run when the grenade explosion killed them. Minutes later another enemy grenade rolled near the group of drivers. Sellman kicked it off the road behind him. After it exploded, another enemy grenade rolled under Seay's trailer approximately three meters from his position. Again Seay left his covered position and threw the armed grenade back at the enemy. At the same time Sellman shot an enemy soldier crawling through the fence. After returning to his position, Seay and Sellman killed two more enemy soldiers trying to crawl through the fence. Suddenly a bullet shattered Seay's right wrist and he told Sellman to cover him while he ran back to the rear of the convoy looking for someone to treat his wound.[351]

Seay located LT Howard Brockbank, SP4 William Hinote, and four other drivers gathered together. Hinote saw Seay had lost much blood and was in pain. One man applied a sterile dressing on the wound, but it did not stop the bleeding. Hinote then tied a tourniquet around Seay's wrist with his shirt. In spite of his wounds, Seay continued to give encouragement and direction to his fellow soldiers. Hinote mentioned his concern about Seay's shattered wrist.

LT Larry Ondic on the 48th Group's first experiment with gun trucks prior to the ambush that killed Seay. Photo courtesy of Larry Ondic

Seay told him to stay alive and not worry about him. One soldier fired a full clip of his M-16 in one burst and Seay admonished him, "Take it easy! Don't waste your ammo - we may run out. What will we do then, stand up and fight them with our fists? I wouldn't be any good at that!" Weak from the loss of blood, Seay moved to the relative cover of a shallow ditch to rest.[352]

For 30 minutes the ambush continued without reinforcement until at 1215 hours, when an MP in the second serial located an undamaged vehicle-mounted radio and called for support. C Company, 4/23rd Infantry immediately notified their brigade headquarters of the ambush. One platoon with two APCs guarding the road one kilometer to the south hurried to the kill zone as soon as they heard the ambush, and was halted by enemy fire from the length of the ditch and a farmhouse 200 meters away. Another platoon of four APCs five kilometers north likewise raced to the kill zone and also came under intense rocket and small arms fire from a Buddhist temple. The enemy made another rush and began crawling over the trucks unloading its cargo. About 1220 hours, COL Wolf arrived over the ambush area in his command helicopter, "Little Bear" 120 from A Company, 25th Aviation; and lacking an immediate reaction force, he ordered C Company, 3rd Battalion, 22nd Infantry, which was involved in a search and attack mission, to air assault into the northern end of the kill zone. At 1235 hours, ten helicopters lifted the infantry company and delivered it five kilometers north of the Buddhist temple at around 1300 hours. The dismounted infantry then accompanied the four APCs toward the village. They encountered civilians fleeing the village, and then discovered dead civilians caught in the crossfire and drivers killed in their cabs. They reached the ARVN compound where they came under mortar fire. Medics, SP4 Ivan Kratzenmeier and Daniel Orozco, quickly established a casualty collection point where they treated wounded flooding into their location.[353]

A squad from the 65th Engineer Battalion, led by SGT Gregory Haley and accompanied by two APCs from the 1st Battalion, 5th Infantry, happened to be sweeping the road for mines and came upon the rear of the convoy. The APCs could not drive past the burning ammunition trailers so SP4 Sellman and other drivers took up a fighting position behind the APC. One of the .50 caliber machine guns on their APC was burned out from previous fighting and the other jammed continually. So the squad engaged the enemy with M-60s, M-16s and grenades. As ammunition began to run low, SGT Haley maneuvered to the rear of one of the APCs and secured more ammunition and another M-60. He jumped down from the APC only to realize his weapon had no trigger. He returned and grabbed another, fed a belt of ammunition into the M-60, and opened fire. The weapon jammed. He pulled the charging handle and it broke off in his hand. As the battle progressed, Sellman was also wounded by shrapnel.[354]

Both ends of the kill zone became casualty collection points, and as the battle raged drivers worked their way to casualty collection points at either end of the convoy. The north end came under mortar fire and Kratzenmeier had the wounded moved further away until he too was wounded by shrapnel. The APCs served as ambulances taking wounded back to the landing zone where helicopters evacuated wounded throughout the afternoon.[355]

After another half hour of fighting, Hinote brought Seay some water. They occasionally fired at enemy positions while waiting for the next attack. Seay noticed three enemy soldiers who had crossed the road to fire at his comrades. Seay raised to a half crouch and fired his rifle with his left hand, killing all three. Suddenly, a sniper's bullet struck Seay in the head, killing him instantly. He only had 60 days left in country.[356]

At 1305 hours, Division G-3 offered an infantry battalion and COL Wolf released ten helicopters to pick them up, but for some unexplained reason the infantry never arrived and the helicopters were not released by Division. At 1310 hours, Troop B, 3rd Squadron, 4th Cavalry landed on the south end of the kill zone and was directed to attack the enemy in the farmhouse 200 meters south of the rubber plantation. During the 20-minute fight, the commander and four men were killed and 11 wounded, but they drove back the enemy. COL Wolf then directed the acting commander to leave one platoon in place and pursue the enemy into the rubber plantation. Taking the farmhouse had freed the two APCs from C, 4/23rd to join the platoon while the rest of Troop B maneuvered to 100 meters east of the Buddhist temple. By 1430 hours, Division moved the 155mm battery within range to provide fire support, which began to turn the tide of the battle. Five tractors and a gun jeep that had reached Tay Ninh dropped their trailers and returned by a back road to help recover damaged vehicles and trailers.[357]

By 1450 hours, the ceiling lifted marginally and two Huey UH-1C helicopters, equipped with two door gunners, 14 rockets, and a mini-gun, from B Company, 25th Aviation Battalion, "Diamondhead," responded first. CWO Robert J. "Bob" Spitler was one of the first to arrive. The commander on the ground informed him the enemy was in the rubber plantation bordering the road. Spitler identified Americans in the ditch and enemy soldiers unloading the trucks. They carried the supplies into the tree line where they were loaded onto their own trucks. The low ceiling, however, prevented the helicopters from attacking at regular angles. The Hueys normally rolled in on the target from a steep dive from about 1,500 feet. The pilots instead had to fly in above the tree tops and fire their rockets on a flat trajectory at point-blank range, all the while receiving enemy ground fire. After expending most of their fuel and ammunition, they hovered low over the tree line to save fuel and simultaneously fired rockets, door guns and mini-guns at the enemy. The enemy was everywhere. Soon, the pilots ran out of ammunition and called for the next "Diamondhead" light-fire team to replace them. Spitler briefed them in the air and the transition of battle went seamlessly. The two Hueys flew back to Cu Chi to refuel, rearm, and wait for the next mission. Those gunships kept the enemy at bay.[358]

By 1500 hours, C Company, 3/22nd Infantry and 4th Platoon, C Company, 4/23rd Infantry were in position to the north to suppress the Buddhist temple while Troop B assaulted it from the rear driving away over 100 enemy into the trench line. By 1530 hours, COL Wolf ordered the forces to roll up the enemy in both the village and rubber plantation, but even with the help of gunships from Diamondhead, the two depleted companies reported the enemy too strong to drive out of the trench. By 1700 hours, the two company commanders were low on ammunition and requested permission to withdraw and resupply for another attack. As the force withdrew a storm hit grounding all the helicopters. About 1830 hours, the defenders developed a plan to drive vehicles out of the kill zone, but the intensity of enemy fire prevented that. So between 1900 and 2300 hours APCs ferried drivers, the wounded and the dead south to the ARVN compound at Go Dau Ha. Those MPs and drivers who had not reached the

7th Transportation Battalion tractors with a fabricated machinegun mount. Photo courtesy of US Army Transportation Museum

safety of the APCs defended in small pockets scattered along the convoy.[359]

By 2205 hours, the defenders sent an emergency call for ammunition, and a helicopter from A Company, 25th Aviation decided to attempt the resupply and came in low under the ceiling to deliver the needed ammunition. It took out more wounded and then the weather began to clear, so helicopters continued missions throughout the night. By the next morning, the relief column finally arrived and the clear sky allowed air support, which drove the enemy away by 1000 hours.[360]

Out of 51 drivers caught in the kill zone, seven lost their lives, ten more were wounded and two, SSG Kenneth R. Gregory and SP4 Bobby L. Johnson both of the 62nd Transportation Company, were taken prisoner. Gregory later escaped and wondered around the jungle for four days before being rescued. Johnson was released five years later in February 1973. The 720th MP Battalion had one man killed, six wounded, lost all six gun jeeps and radios, and had five M-60s destroyed. The relief force lost 23 killed and 35 wounded. SP4 Seay posthumously received the Medal of Honor and was promoted to sergeant. SP4 Sellman received the Bronze Star Medal with V device for his valor. SSG Frazier and SP4 Murphy, 720th MP Battalion, both received Silver Star Medals and SGT Raymond H. Tate the Bronze Star Medal. SP4 Guy A. Davison, of the 720th MP Battalion, posthumously received the Silver Star Medal. SP4 Kratzmeier received the Bronze Star Medal for his bravery in treating and evacuating wounded under fire.[361]

SGT William Seay.

Lesson

The unusual aspect of this ambush was the size of the ambushing force and the effort the enemy made to unload cargo from the vehicles. Normally the enemy just wanted to inflict damage in the convoy, but in this case they seemed very interested in the cargo, so they brought a large enough force that could beat back the local security forces. MG Williamson's decision to pull the 3rd Brigade back to Saigon allowed the 88th NVA Regiment greater freedom of movement.

This was the first major convoy ambush for the 48th Group in the III Corps Tactical Zone and it had relied entirely on the combat arms units for protection. This external support was subject to the decision of the commander based on competing priorities. The arbitrary decision of the area commander to defend Saigon stripped the area of combat arms support and required the reaction force to come from a greater distance. In the absence of gun trucks or combat arms, the drivers had to fight as infantry men. They needed to know all the skills for setting up a defense and coordinating fire support.

Since the small security forces could not fight through the kill zone, they established defensive positions on both ends of the kill zone and these became the casualty evacuation sites and rallying points for drivers who worked their way to them.

Action

Having faced a devastating convoy ambush similar to what 8th Group experienced on 2 September, the 48th Transportation Group came up with a different solution to ambushes than the 8th Group. The 48th Group started by making everyone wear their helmets

Vietnam War 141

and protective vests. Not suffering anyone killed in an ambush since 22 November 1966, the drivers had become complacent and quit wearing protective gear because of the heat. The 48th Group also required that trucks include assistant drivers as "shotguns." This added extra riflemen in a fight. In previous ambushes, they had followed the SOP to not stop in the kill zone, but they had no choice if vehicles blocked the narrow roads as on 25 August. They could not turn around and drive out of the kill zone. In this case, the drivers had to fight as infantry until the nearest reaction force arrived. The slow response by the reaction force on 25 August was an embarrassment for the 25th Infantry Division, so the soldiers set out to resolve the problem.[362]

In August 1968, representatives from the MP, division, and transportation units held several conferences to define relationships. According to a report in 1971, the Provost Marshal of the 25th Infantry Division assumed responsibility for convoy security for the 48th Group convoys. He flew overhead in an aircraft and shared control of the convoy with the convoy commander on the ground. In the event of an ambush, infantry or cavalry commanders assumed control of the convoy.[363]

COL Paul Swanson assumed command of the 48th Group in November 1968 and opposed the use of gun trucks. This was surprising since the 6th Battalion had previously built gun trucks to conduct a night convoy south to support the 9th Infantry Division in the fall of 1967. Anticipating ambushes, the 6th Battalion welded steel plates to the doors of 20 light trucks and two jeeps. They also fabricated machine gun mounts and welded them to the right side of the cab so that the assistant drivers could fire them while standing. The convoy was ambushed and the gun trucks successfully repelled the attackers. The battalion conducted no more night convoys and the idea of gun trucks spread no further through the 48th Group.[364]

Instead, Swanson believed the combat commander had the responsibility for convoy security. The ambushes usually ended when the infantry and tanks arrived and swept through the area. He did not want to crowd into the infantry's mission or take task vehicles off the road. Swanson did, however, allow drivers to put steel plating on the sides of their cabs for individual protection. The main issue with the ambush at Ap Nhi was that the field commander arbitrarily pulled the infantry from defending the road to defending Saigon leaving the convoys vulnerable. The trucks needed some guarantee the combat arms would not leave them unprotected again. So Swanson told the 1st and 25th Infantry Divisions that if they wanted their beer and soda, they needed to keep the enemy off of his convoys. It helped that the G-3 of the 25th Infantry Division was a classmate of his from the Army War College. For the next year, the 25th Infantry Division provided excellent support. This relationship was all personality-driven though.[365]

---

[343] Stephen C. Tunnell, "Convoy Ambush at Ap Nhi," Vietnam, April 1999; and August 1968 Battalion Time Line, 720th Military Police Battalion Reunion Association Vietnam History Project, http://720mpreunion.org/history/time_line/1968/08_1968.html.
[344] Tunnell, "Convoy Ambush."
[345] COL Frank B. Case, Operational Report for Quarterly Period Ending 31 October 1968, Headquarters, 48th Transportation Group (Motor Transport), 8 November 1968; and Ivan Katzenmeier, "A Combat Medic's Vietnam Experience," Katzenmeier's Weblog, A Journey in Words and Photos, http://katzenmeier.wordpress.com/.
[346] Tunnell, "Convoy Ambush;" Ron Leonard, "The Ambush At Ap Nhi," http://25thaviation.org/id804.htm; and narrative by Marvin E. Branch, Katzenmeier's Weblog.
[347] Case, Operational Report; Tunnell, "Convoy Ambush;" Leonard, "The Ambush At Ap Nhi;" and 720th MP time Line.
[348] Case, Operational Report; Tunnell, "Convoy Ambush;" and Katzenmeier, "A Combat Medic's Vietnam Experience."
[349] 720th MP time Line.
[350] Tunnell, "Convoy Ambush at Ap Nhi;" and Katzenmeier, "A Combat Medic's Vietnam Experience."
[351] Tunnell, "Ambush at Ap Nhi;" and SGT William Seay Medal of Honor Citation.
[352] Tunnell, "Ambush at Ap Nhi."
[353] Case, Operational Report; Katzenmeier's "A Combat Medic's Vietnam Experience."
[354] Tunnell, "Convoy Ambush;" and narrative by Marvin E. Branch, Katzenmeier's "A Combat Medic's Vietnam Experience."
[355] Katzenmeier's "A Combat Medic's Vietnam Experience."
[356] Tunnell, "Ambush at Ap Nhi;" and Seay Medal of Honor Citation.
[357] Tunnell, "Ambush at Ap Nhi;" and Leonard, "The Ambush At Ap Nhi."
[358] Leonard, "The Ambush At Ap Nhi.."
[359] Case, Operational Report; and Leonard, "The Ambush At Ap Nhi.."
[360] Leonard, "The Ambush At Ap Nhi.."
[361] 720th MP Battalion Time Line.
[362] BG Orvil C. Metheny Interview by CPT Louis C. Johnson, 22 November 1985; BG (R) Orvil Metheny summary of telephone interview by Richard Killblane, 19 March 2004; and LTC Orvil Metheny, Operational Report – Lessons Learned, Headquarters, 6th Transportation Battalion (Truck), Period Ending 31 October 1968, 20 February 1969.
[363] Metheny interviews and Thomas, Vehicle Convoy Security," p. II-32.
[364] Discussion with LTC (R) Larry Ondic.
[365] Metheny interviews.

Conclusion

Check Point 88, intersection of QL19 and the Pleiku bypass, where convoys dropped off their gun trucks and headed to their destinations. An MP V100 is in the foreground. The photo was taken in late 1969 since True Grit is parked on the right and Cold Sweat above it has the second paint scheme. *Photo courtesy of US Army Transportation Museum*

Convoy operations during the Korean and Vietnam Wars revealed certain lessons. When stopped in a kill zone without the benefit of gun trucks, the only recourse is to dismount and return fire. The aggressive response would also include fighting through the enemy as infantry. This is supposing the truck drivers have been trained adequately as infantry. An inherent problem is the spacing of the vehicles even when bunched up probably does not allow a concentration of truck drivers for a successful assault thus violating the principle of mass. So truck drivers came up with a different response.

Truck drivers during both the Korean and Vietnam War realized the truck, if mounted with a crew-served weapon, provided both maneuverability and firepower for the fight inside the kill zone. The kill zone is a place if planned properly of death and destruction, so it is imperative to exit it as fast as possible. The infantry and armor option is to fight through the enemy, but truck drivers can drive out. Wheeled vehicles provide the advantage of speed and can also serve as a weapons platform. During the Korean War, the gun truck ratio was 1:3, but the ring-mounted cargo trucks hauled cargo like any other truck in the convoy so they employed the passive tactics of returning fire and clearing the kill zone. During the Vietnam War, the truck drivers added armor to the trucks making them dedicated fighting platforms thus allowing them the freedom to maneuver inside the kill zone to defend disabled vehicles and rescue any drivers. The gun truck ratio during the Vietnam War was one gun truck per ten task vehicles, but each gun truck carried at least three crew-served weapons resulting in a weapon ratio of nearly 1:3 as during the Korean War. Over time, the gun truck crews learned to add a variety of weapons for contingencies such as the enemy getting right up next to the gun truck. The gun truck of the Vietnam War was a true fighting platform capable of maneuvering throughout the kill zone.

Vietnam War

Another lesson learned, forgotten and relearned during the 19th century and Vietnam War was a convoy march serial should not exceed 30 vehicles. 30 vehicles with a 100-meter interval stretches out to three kilometers, or a mile. Convoy commanders admit it is much easier to control smaller convoys of this size and the gun truck ratio of 1:10 allows for two to three gun trucks in the convoy serial.

The mission of the convoy is to deliver the goods to the customer and the mission of the gun trucks is to deter enemy attacks, but if attacked to defend the convoy until the nearest combat force can arrive and finish the fight. During the Vietnam War, the arrival of helicopter gunships usually ended any ambush, and it did not take long to learn the presence of gunships likewise deterred enemy ambushes.

For a gun truck to coordinate its maneuver and those of other gun trucks, the ground combatants and close air support, it needed radios – one of the three tenants of ground combat; shoot, move and communicate. Radios allow the participants to synchronize the battle. While the addition of any crew-served weapon turned a truck into a gun truck, the feature that made Vietnam gun trucks stand out was its armor and configuration of weapons. The Vietnam gun truck became dedicated fighting platform.

As much as was learned, the Vietnam War left three unresolved issues concerning convoy security. The first of which dealt with the type of escort platform. The 8th Transportation Group experimented with different types of armor and armament on nearly every type of wheeled vehicle in its inventory. The initial escort platform was the gun jeep with the single M-60 machine gun. The M151 ¼-ton trucks did not have the power to carry much armor and were too small to mount more than one machine gun, but had room for several sets of radios. The M37 ¾-ton could carry the armor and also mount two M-60s. Consequently, these two vehicles made ideal command and control vehicles for convoy and assistant convoy commanders.

During the first attempt to convert cargo trucks into gun trucks, the crews constructed protective walls of sandbag and wood but realized the rough conditions of the road beat the sandbags apart, additionally they were heavy and in a firefight, sandbags tended to leak out their contents when hit. ¼-inch steel plates around the bed and on the doors of 2½-ton trucks turned them into more effective fighting platforms, but single walls only stopped small arms fire, not 12.7mm or RPGs. The M35 2½-ton truck did not have the power to hold more weight and keep up with the task vehicles. The M54 5-ton cargo truck not only had the power to accelerate past the task vehicles but could do so with a second steel wall. By the end of the war, the 5-ton with double steel wall became the preferred choice for gun truck.

Due to the shortage of steel plates, both the 8th Transportation Group out of Qui Nhon and the 500th Group down in Cam Ranh Bay experimented with mounting M113 APC hulls on the backs of 5-ton cargo trucks. These types of gun trucks proved top heavy and dangerous to drive especially when making turns. The 8th Group crews slid their APC hulls in front forward which placed most of the weight over the back wheels, but the 500th Group crews slide their APC hulls in backwards which centered the weight better between the front and rear wheels.

During the Korean War, the truck drivers mounted machine guns on ring mounts and found the M-2 .50 caliber was more intimidating than the .30, just as the Vietnam veterans learned the superior fire power of the .50 over the 7.62mm M-60. The 2½-ton gun trucks started out with two M-60 machine guns but graduated to three to four .50 caliber machine guns arrayed around the gun box so they could provide spraying fire across 360 degrees. Some gun trucks even acquired M-134 mini-guns, which put out tremendous fire power but the rough road could knock the timing off. During the first year of experimentation, the 8th Transportation Group borrowed Quad .50s mounted on 2½-ton trucks. They threw out a lot of lead, but required a crew of six; driver, gunner and four loaders. Since the table of organization and equipment (TOE) did not authorize gun trucks, it did not provide the additional personnel to man them; so the truck companies had to take drivers off task vehicles to man the gun trucks. Each crew member of a gun truck represented a task vehicle that did not have a driver. The Quad .50s were not cost effective and did not provide any armor for the loaders. The Quad .50s could not depress to engage enemy on the sloping terrain in most mountain passes and did not offer 360 degrees of fire since the cab of the truck got in the way. Consequently, the M-2 .50 caliber machine gun designed by John Browning in 1918 proved the preferred weapon of choice and only the number and location on the gun truck differed.

Even COL Bellino considered the gun truck a field expedient until enough Cadillac Gage V100 four wheeled armored cars arrived to replace them. The delay in Transportation units receiving the V100 was MPs had priority for issue of the V100. Even MG Joseph M. Heiser, Jr., Commander of 1st Logistics Command, considered the V100 a more effective escort vehicle and forwarded his recommendations in a report to the Assistant Chief of Staff for Force Development in August 1969. Every senior officer in the chain of command concurred; however, the gun truck crews knew its weaknesses. While the senior officers saw the V100 fully enclosed armored car as offering 360 degrees of armor protection, the gun truckers realized there was no place for the blast of an RPG to escape if it penetrated the armor. Unless the RPG penetrated right where the gunner in a gun box was standing he would live, but the armored car trapped the blast inside which would kill or seriously wound the entire crew. These case studies presented several cases where gunners such as John Dodd or Chester Israel survived RPG hits in 5-ton gun boxes and continued to fight. The V100 only had a single cupola with two M-73 7.62mm machine guns that could only fire in one direction, where the 5-ton gun truck could fire in three directions at once. Not only that but the V100 did not have the power train that could keep up with the convoy. A later report by an Army concept team conducted from December 1970 to March 1971 agreed with the truck drivers.[366]

8th Transportation Group agreed that gun truck crews should remain within their assigned companies so at the end of the day they had to bunk with the men they were charged to protect. The role of the gun trucks was to go into the kill zone while everyone else was trying to get out of it, so not every truck driver rose to that challenge. In the beginning, commanders appointed their best drivers to the gun trucks but over time, the gun truck gained an élan, mostly due to the elaborate art work, and the crews selected their replacements. The 500th Group in Cam Ranh Bay did likewise but only assigned a permanent crew of two, driver and gunner, to minimize the reduction of drivers of cargo trucks. They pulled the other two gunners from a pool of volunteers drawn from the depot units at Cam Ranh Bay. This left no continuity unless the gunner volunteered every day. The 48th Group down in Long Binh avoided gun trucks up until end of the war, and then consolidated all their gun trucks into a single truck company. The advantage of consolidating the gun trucks in one company is simplicity in mission, training and maintenance. Both gun trucks internal and external to the truck companies offered advantages and disadvantages that would resurface during the next war requiring gun trucks.

Tactics describes the way battles are fought. The gun truck crews of the 8th Group had only one tactic – to drive into the kill zone spraying in every direction until either they ran out of ammunition or the ground security force or helicopter gunships arrived. The 8th Group Standard Operating Procedures (SOP) called for the gun trucks to move to the edge of the kill zone to lay down suppressive fire, but most gun trucks drove into the kill zone especially where any disabled cargo trucks might be. The edge of the kill zone provided a safe distance from enemy fire while their .50s could still lay down effective suppressive fire. Since the muzzle flash did not betray the enemy's position during the day, the gun truck gunners fired in any direction hoping for chance to hit the enemy. A study of the history of infantry tactics reveals that ground is controlled by a given amount ammunition expended in a specific amount of time. For example a Civil War rifle company of a hundred men could fire two to three rounds per minute, but a rifle squad during World War I occupied the same about of ground. The bolt action rifle increased the rate of fire for ten riflemen to 10 to 15 rounds per minute and the addition of either 250 rounds per minute of the Chauchat or 500 to 650 rounds per minute of the Browning Automatic Rifle (BAR). The actual rate of fire of the BAR was significantly reduced by the fact the gunner and assistant gunner only carried 12 20-round magazines each. So without accurate fire, 200 to 300 rounds controlled the same amount of ground by the random probability of hitting something. Without knowing exactly where the enemy was hidden, the M-2 .50 caliber machine gun could send 450 to 550 rounds down range in one minute and over time the gun trucks carried enough ammunition cans to fire the full capability of the weapons.

The gun truck crews of the 500th Group employed a technique of what Iraqi gun truck crews would call "bumping." They ran with long convoys and the gun truck in the kill zone would lay down a base of fire employing four to six round bursts until the next gun truck arrived. The next gun truck would take the first's place and it would then continue with the convoy out of the kill zone. Each successive gun truck would come up and take the place of the one before it until every cargo truck had cleared the kill zone.

Another unresolved issue was whether the aggressive tactics of 8th Group or the passive tactics of the 500th Group worked best. The passive tactics referred to returning fire and clearing the kill zone. The aggressive tactics of the 8th Group called for every gun truck that could reach the kill zone to concentrate in it and tear up anything within range of their weapons.

The First World War heralded the beginning of the broad front concept of war that did not end until Operation Desert Storm. During World War I the Allies dug trenches that connected the English Channel with the Swiss Alps creating a front line with the enemy before it and a safe rear behind it. The lack of a front line during the Vietnam War made this conflict seem as an anomaly of the 20th century rather than the rule, so the lessons of convoy security were forgotten as the Army tried to distance itself from the loss and refocus on Fulda Gap and the possible war with the Soviet Union. Not until the war in Iraq in 2003 would the US Army realize convoy security needed to be a permanent part of doctrine.

---

[366] Richard E. Killblane, Circle the Wagons; The History of US Army Convoy Security, Ft Leavenworth, KS: Combat Studies Institute, 2005.

Logan Werth and Mike Cameric on Uncle Meat giving the thumbs up because they had survived another ambush. Besides covering the floor with ammunition cans, they had stacked more in the corners because of the increase in ambushes during Lam Son. Photo courtesy of Logan Werth

8th Transportation Group SOP Annex A 1968 and 1969:

8. Action to be taken in the event of ambush, sniper fire or mines:

a. *Ambushes:*

When a convoy encounters sniper fire, an ambush or mines it should KEEP MOVING if at all possible.

The security forces will be contacted by radio immediately.

All vehicles which are beyond the kill zone will continue moving toward their destination.

If unavoidably stopped in the kill zone, personnel must dismount, take up covered and concealed positions, and lay down a heavy volume of return fire only on active enemy targets, staying in close proximity of their vehicles. In no case will convoy personnel close with or attempt to pursue the enemy as this will interfere with the tactical reaction plan. Convoy personnel will cease fire upon arrival of tactical security elements to allow tactical vehicles and personnel freedom of movement against enemy forces. Convoy control personnel will insure that traffic begins to move as soon as the tactical situation permits.

Those vehicles which have not yet entered the kill zone will halt at 100 meter intervals, dismount and defend as described in the preceding paragraph, firing only if the enemy targets are visible.

If the tactical situation permits move disabled vehicles off the road.

b. *Mining Incidents:*

If mines are encountered, the convoy commander must, in conjunction with the tactical commander, quickly determine a course of action. While a single mine should not delay an entire convoy, vehicles must not enter an area known to be heavily mined.

If the convoy is halted by mines, the convoy commander will immediately alert the security forces and disperse his personnel around the vehicles in preparation for an attack.

c. *Sniper fire:*

If the convoy or a segment of the convoy receives sniper fire, fire should be returned and vehicles will continue moving.

The convoy commander will attempt to determine the location of the sniper and report it to the tactical forces securing that portion of the route.

9. General Precautionary Measures:

Good convoy discipline and continuous driver training will reduce the possibility and/or effectiveness of enemy action against convoys.

Drivers should track the vehicle ahead when these vehicles are in sight. This will reduce the possibility of detonating a pressure-activated mine which the vehicle ahead may have missed. Conversely, old tracks should be avoided because Viet Cong commonly places mines in old tracks.

Speed generates carelessness. The enemy relies on careless drivers in his employment of mines.

Key personnel, who are prime targets for command detonated mines, must avoid congregating in one vehicle or location within the convoy.

Drivers should avoid carelessly driving over wooden sticks or other debris on the road. Pieces of wood (board or bamboo) on a roadway may indicate the presence of pressure activated mines or other explosive devices.

Vehicle operators should be alert for changes in familiar scenes, evidence of apparent road repairs, new fill or paving, road patches, mud smears, grass, dirt, dung or other substance on roads. Such areas could be evidence of enemy

Each convoy commander will make such coordination as necessary with escorts or at critical points in sufficient time to prevent any delay of the convoy. Where unforeseen delays or route changes occur, the convoy commander will expeditiously notify his Battalion headquarters which will pass this information to the Group Operations section.

Information will include the cause, location (grid coordinates), alternate routes available and other pertinent data.

8th Transportation Group Tactical Convoy Operations, 1969:

14 points of ambush:

Keep weapon ready to fire.

Be alert for changes in familiar scenes along route.

Use your weapon, return fire immediately.

Notify security forces by radio, call – contact, contact, contact.

Continue moving, maintaining interval if possible.

Track vehicles ahead to prevent further detonation of mine.

Don't enter kill zone if possible.

If disabled and convoy is moving through kill zone, mount a passing vehicle.

Provide flanking support fire into kill zone.

Gun trucks apply maximum base of fire. Fight as a team. Deploy upon command.

Prime targets are: enemy automatic weapons positions, enemy mortar, rocket positions and assaulting ground troops.

Stay on hard surfaces.

Cease fire on signal.

Remember details.

Bibliography

Interviews

Baird, James
Ballard, Larry "Ted"
Barrowcliff, Merton
Belcastro, Dennis
Bennett, Calvin
Blosser, Paul
Briggs, Thomas
Brown Phillip C.
Buckwalter, Jack
Buirge, Michael
Burke, COL (R) John
Cahill, Richard
Calhoun, J.D.
Calibro, Steve
Carter, Fred
Champ, Roger D.
Chapman, John
Chalker, Wayne C.
Cochran, Danny
Collins, LTC (R) Nicholas
Darby, Jim
Deeks, Walter
Dalton, Robert
Dobos, Wayne
Dodd, John
Dye, Robert
Eichenberg, LTC (R) William L.
Fiandt, Larry
Freeman, Erik
Fuller, CPT (R) Ralph
Gress, John
Hatton, Alford

Honor, LTG (R) Edward
Israel, Chester
Jacobs, John
Kendal, Ron
Looney, Gary
Lyles, James
Mallory, Ronald
Marcum, Paul
Marshall, Harold
Mauricio, Paul
McCarthy, Joseph
Medley, Walter Dan
Metheny, BG (R) Orvil C.
Montgomery, Barry
Ondic, LTC (R) Larry
Owens, Ron
Patrick, Wayne
Robertson, Walter
Seay, Sammy
Sims, Charles
Smiley, COL (R) Philip N.
Smith, Rick
Smith, Ronald
Swanson, COL (R) Paul
Taylor, Clifford Wesley
Tellez, Manny
Van Brocklin, Terrance
Voightritter, LTC (R) Ronald
Weeks, Grailin
Welton, Burrell
Wheat, Timothy
Wolfe, COL (R) Melvin M.

Papers and Reports

1st Marine Division Fleet Marine Force Special Action Report for the Wonsan – Hamhung – Chosin Reservoir Operation; 8 October – 15 December 1950, Volume One, May 21, 1951.
After Action Report to Commander C Company, 504th MP Battalion.
"Ambushes in the An Khe Pass," unknown author of the 27th Transportation Battalion, 1970.
Appleman, LTC Roy E. Papers, US Army Historical and Education Center.
Award Citations for SP4 Ronald M. Kendall and SP4 Timothy Wheat, 7 February 1968.
Barrett, MAJ Jonathan R., Operational Report of the 36th Transportation Battalion (Truck) for Period Ending 31 January 1969, 12 February 1969, RCS CSFOR-65 (RI).
Bing, LTC Tom L., Operational Report of the 124th Transportation Battalion (Truck) for the Period Ending 31 January 1969, 4 February 1969 CSFOR-65 (R-1).
____, Operational Report of the 124th Transportation Battalion (Truck) for the Period Ending 30 April 1969, 8 May 1969 CSFOR-65 (R-1).
Bozeman, LTC Wallace B., Operational Report of the 57th Transportation Battalion (Truck) for Quarterly period Ending 31 October 1969, 9 November 1969. Report Control Symbol CSFOR-65 (R-1).
Brown, COL Albert J., Trip Report (8-69). Convoy Security, US Army Combat Developments Command, Liaison Detachment, HQ USARV, APO San Francisco, 96375, 20 January 1969.
Busbey, COL Arthur B., Jr., Subject: Hwy 19 SOP for Logistical and Security Elements, Headquarters I Field Force Vietnam, APO San Francisco 96350, 17 July 1968.
Case, COL Frank B., Operational Report for Quarterly Period Ending 31 October 1968, Headquarters, 48th Transportation Group (Motor Transport), 8 November 1968.
Collins, MAJ Nicholas H., "Battalion S-3 Notes," Headquarters, 54th Transportation Battalion, APO 96238, 5 March 1967.
Daniels, LTC Walter C., Operational Report - Lessons Learned (Headquarters, 27th Transportation Battalion, Truck), for Period Ending 31 October 1969, 13 November 1969.
_____, Operational Report - Lessons Learned (Headquarters, 27th Transportation Battalion, Truck), for Period Ending 30 April 1970, 13 May 1970.
_____, Operational Report - Lessons Learned (Headquarters, 27th Transportation Battalion, Truck), for Period Ending 31 July 1970, 13 August 1970.
Department of the Army, Headquarters, US Army Support Command, Qui Nhon, General Order Number 102, Award of the Bronze Start Medal for Heroism, 2LT Burrell Welton, 7 February 1968.
Descoteau, 1LT Roland E., Unit History, 88th Transportation Company, 1 January to 31 December 1967, 25 March 1968.
Durant, LTC John J., Operational Report - Lessons Learned (Headquarters, 27th Transportation Battalion, Truck), for Period Ending 31 October 1970, 14 November 1971.
Edmiston, MAJ Charles H., Operational Report of the 57th Transportation Battalion (Truck) for Quarterly period Ending 30 April 1969, 3 May 1969. Report Control Symbol CSFOR-65 (R-1).
Ellis, LTC Alvin C., "Operational Report – Lessons Learned, 39th Transportation Battalion (Truck), Period Ending 30 April 1971, RCS CSFOR-65 (R2)," Headquarters, 39th Transportation Battalion (Truck), APO San Francisco 96308, 4 May 1971.
Fitzgibbons, LTC Eugene T., Operational Report of the 36th Transportation Battalion (Truck) for Period Ending 31 July 1968, 31 July 1968, RCS CSFOR-65 (RI).
Forster, CPT Paul, Unit History, Annual Supplement, 512th Transportation Company (LT TRK), 1 January to 31 December 1967, 18 March 1968.
Freeman, Erik and Charles Sims email correspondence, 15 – 21 February 1998.
Freeman, LTC William C., Operational Report of the 27th Transportation Battalion (Truck) for the Period Ending 30 April 1970, 6 May 1969.
_____, Operational Report of the 27th Transportation Battalion (Truck) for the Period Ending 31 July 1969, 1 August 1969.
Gaudio, 1LT Alan N., Operational Report 124th Transportation Battalion (Truck) for period Ending 31 October 1969, 11 November 1970, RCS CSFOR-65 (R-1).
Grabia, 1LT Stanley, Unit History, 666th Transportation Company (Light Truck), C.Y. 1970, 15 April 1971.
Hall, CPT Phillip T., Jr., 585th Transportation Company (Medium Truck Cargo) APO 96238, After Action Report, 11 November 1968.
Hammel, 1LT William C., Unit History, 669th Transportation Company (LT TRK), 1 January 1968 – 31 December 1968, 11 March 1969.
Henderson, LTC Chester T., Operational Report of the 57th Transportation Battalion (Truck) for Quarterly period Ending 30 April 1968, 30 April 1968. Report Control Symbol CSFOR-65 (R-1).
Honor, LTC Edward, Operational Report – Lessons Learned, Headquarters, 36th Transportation Battalion, Period Ending 31 October 1969, 25 February 1970.
Hume, BG A. G., Deputy Chief of Staff (P&O), Combat Lessons Bulletin Number 14, Headquarters, US Army Vietnam, 20 October 1970.
Hunt, 1LT James F., Operational Report – Lessons Learned, Headquarters, 500th Transportation Group, Period Ending 30 April 1969, 5 September 1969.
Jenkins, 1LT J. S. B., Unit History Annual Historical Supplement, 512th Transportation Company (LT TRK), 1966, 21 July 1967.
Johnson III, 1LT Harris T., Unit History, 523rd Transportation Company (LT TRK), 1 January to 31 December 1968, 31 January 1969.
Klein, 1LT Ronald F., Unit History, 666th Transportation Company (Light Truck), C.Y. 1968, 15 March 1969.
Kramer, LTC John C., Operational Report 124th Transportation Battalion (Truck) for period Ending 31 July 1969, 8 August 1969 (RCS CSFOR-65) (R-1).
_____, Operational Report 124th Transportation Battalion (Truck) for period Ending 31 October 1969, 8 November 1969 RCS CSFOR-65 (R-1).
Langston, COL Alex T., Jr., Operational Report of the 8th Transportation Group (Motor Transport) for the period ending 31 January 1970, 15 February 1970 RCS CSFOR-65 (R-1).
_____, Operational Report of the 8th Transportation Group (Motor Transport) for the period ending 30 April 1970, 22 May 1970 RCS CSFOR-65 (R-1).
_____, Operational Report of the 8th Transportation Group (Motor Transport) for the period ending 31 July 1970, 19 August 1970 RCS CSFOR-65 (R-1).
Little, LTC C. D. M., Operational Report of the 57th Transportation Battalion (Truck) for Quarterly Period Ending 31 July 1970, 6 August 1970. Report Control Symbol CSFOR-65 (R-1).
Ludy, COL Garland A., Operational Report of the 8th Transportation Group (Motor Transport) period ending 31 July 1969, 10 August 1969.
_____, Operational Report of the 8th Transportation Group (Motor Transport) for the period ending 31 July 1969, 10 August 1969 RCS CSFOR-65 (R-1).

_____, Operational Report of the 8th Transportation Group (Motor Transport) for the period ending 31 January 1969, 7 February 1969 RCS CSFOR-65 (R-1).
Memorial Service, 54th Transportation Battalion, Camp Addison, Republic of Vietnam, 6 September 1967.
Metheny, LTC Orvil, Operational Report – Lessons Learned, Headquarters, 6th Transportation Battalion (Truck), Period Ending 31 October 1968, 20 February 1969.
Nantroup, MAJ John F., Operational Report of the 500th Transportation Group (Motor Transport) for Period ending 31 July 1969, RCS CSFOR-65 (RI), 8 December 1969.
Petro, George E., Appendix 10 to Annex Peter Peter to 1st Marine Division Special Action Report, 30 December 1950.
Preble, 1LT Alvin L., Unit History, 359th Transportation Company (Medium Truck Petroleum), 1 January to 31 December 1968, 6 April 1969.
Puller, L. B., Annex Peter Peter to 1st Marine Division Special Action Report, HQ 1st Marines (Reinf), 15 January 1951.
Rackley, LTC Jerry D., Operational Report of the 54th Transportation Battalion (Truck), WFR6AA, for the Period Ending 31 October 1970, 5 November 1969.
_____, Operational Report - Lessons Learned (Headquarters, 27th Transportation Battalion, Truck), for Period Ending 31 January 1970, 18 February 1970.
_____, Operational Report 124th Transportation Battalion (Truck) for period Ending 30 April 1969, 14 May 1970 (RCS CSFOR-65) (R-1).
Ratcliff, LTC R. H., Operational Report of the 57th Transportation Battalion (Truck) for Quarterly period Ending 30 April 1970, 7 May 1970. Report Control Symbol CSFOR-65 (R-1).
Reise, LTC Paul E., Operational Report of the 36th Transportation Battalion (Truck) for Period Ending 31 October 1968, 31 October 1968, RCS CSFOR-65 (RI).
Sarber, LTC William R., Operational Report of the 54th Transportation Battalion (Truck), WFR6AA, for Period Ending 30 April 1969, 5 May 1969.
_____, Operational Report of the 54th Transportation Battalion (Truck), WFR6AA, for Period Ending 31 July 1969, 8 August 1969.
Shelley, 1LT Jay N., Unit History, 512th Transportation Company (LT TRK), 1 January 1968 – 31 December 1968, 11 March 1968.
Smith, MG Oliver P., Commanding General, 1st Marine Division, Letter to the Commandant of the Marine Corps, 17 December 1950.
Spain, 1LT Donald L., Unit History, 666th Transportation Company (Light Truck), C.Y. 1969, 25 March 1970.
Storey, MAJ Thomas P., Operational Report – Lessons Learned, Headquarters, 36th Transportation Battalion, Period Ending 31 January 1970, Headquarters, 36th Transportation Battalion (Truck), 10 January 1970.
_____, MAJ Thomas P., "Operational Report – Lessons Learned 24th Transportation Battalion (Terminal), Period Ending 31 January 1970, 17 February 1970, RCS CSFOR-65 (R2).
Thomas, COL David H., "Vehicle Convoy Security Operations in the Republic of Vietnam," Active Project No. ACG-78F, US Army Contact Team in Vietnam, APO San Francisco, CA 96384, 30 Sep 71.
Thomas, 1LT Peter C., Unit History Report, 57th Transportation Company, 1 January 1968 – 31 December 1968, 24 March 1969.
Tish, 1LT Thomas L., Annual Historical Summary, 585th Transportation Company (Medium Truck Cargo), 1 January 1968 – 31 December 1968.
Unit History, 523rd Transportation Company (Light Truck), 1 January 1967 – 31 December 1967.
Unit History, 669th Transportation Company (Light Truck), 1 January 1967 – 31 December 1967.
Utley, LTC Jack C., Operational Report – Lessons learned, Headquarters, 54th Transportation Battalion (Trk), Period Ending 30 April 1968, 17 July 1968.
Vitellaro, 1LT John J., Unit History, 669th Transportation Company (LT TRK), 1 January 1969 – 31 December 1969, 25 March 1970.
Weston, Thomas, letter to parents, November 29, 1967 and letter to Squadron Press, August 29, 1996.
Whalen, 1LT Thomas F., Unit History, 512th Transportation Company (LT TRK), 1 January 1969 – 31 December 1969, 1 February 1969.
Wilkins, 1LT William J., Unit History Report, 64th Transportation Company (Medium Truck), 1 January 1968 – 31 December 1968, 7 April 1969.
Wills, 1LT Robert M., Unit History, 523rd Transportation Company (LT TRK), 1 January 1969 to 31 December 1969, 23 March 1970.
Womack, MAJ Daniel, Jr., Operational Report of the 57th Transportation Battalion (Trk) for quarterly period Ending 31 January 1970, 5 February 1970. Report Control Symbol CSFOR-65 (R-1).
Wright, LTC Louie E., Operational Report of the 27th Transportation Battalion (Truck) for the Period Ending 31 January 1970, 3 February 1969.

Books and Articles

Appleman, LTC Roy E., *Escaping the Trap; The US Army X Corps in Northeast Korea*, 1950, College Station, Texas: Texas A&M University Press, 1990.
August 1968 Battalion Time Line, 720th Military Police Battalion Reunion Association Vietnam History Project, http://720mpreunion.org/history/time_line/1968/08_1968.html Belcastro, Dennis, Narrative, "Living History Reenactment at Jamestown March 2000," Army Transportation Association Vietnam, http://grambo.us/belcastro/54guntruck.htm.
Barrowcliff, Merton, "Story," unpublished account of ambush submitted to 359th Transportation Company Vietnam.
Bellino, Colonel Joe O., "8th Transportation Group; Sep 1967 – Sep 1968." n.d.
Chalker, Wayne C., "The Fire Base Bastogne Convoy 4/68," Army Transportation Association Vietnam, 134.198.33.115/atav/default.html.
Dodd, John, "Story," unpublished account of the ambush.
Katzenmeier, Ivan, "A Combat Medic's Vietnam Experience," Katzenmeier's Weblog, A Journey in Words and Photos, http://katzenmeier.wordpress.com/.
Killblane, Richard, *Mentoring and Leading;The Career of Lieutenant General Edward Honor*, Fort Eustis, VA: US Army Transportation School, 2003.
_____, *Circle the Wagons The History of US Army Convoy Security,* Global War on Terrorism: Occasional Paper 13, Fort Leavenworth, KS: Combat Studies Institute Press, 2005.
Leonard, Ron, "The Ambush At Ap Nhi," http://25thaviation.org/id804.htm.
Lyles, James, *Gun Trucks in Vietnam; Have Guns – Will Travel*, Wheaton, ILL: Rhame House Publishers Inc., 2012.
_____, *The Hard Ride; Vietnam Gun Trucks,* Part 2, Quezon City, Philippines: Planet Art, 2003.
_____, *The Hard Ride: Vietnam Gun Trucks;* Part Two, n.p., 2011.
_____, *The Hard Ride; Vietnam Gun Trucks,* Quezon City, Philippines: Planet Art, 2002.
Ludy, COL Garland A., "8th Transportation Group, Sep 1968 – Sep 1969."
McAdams, Frank, *Vietnam Rough Riders; A Convoy Commander's Memoirs,* Lawrence, Kansas: University Press of Kansas, 2013.
Nolan, Keith William, *Into Laos; The Story of Dewey Canyon II/Lam Son 719; Vietnam 1971*, Novato, CA: Presidio Press, 1986.
Tunnell, Stephen C., "Convoy Ambush at Ap Nhi," *Vietnam,* April 1999.

A

Abrams, GEN Creighton, 34
Amos, Marion, 118
Anderson, SP4 George, 39-41, 43-45
Arbuthnot, 1LT John M. II, 30, 49, 50
ARVN units
    1st Squadron, 10th ARVN Cavalry, 92
    1st ARVN Airborne Division, 123
    1st ARVN Armored Division, 123
Almond, LTG Edward M., 8
Aragon, Pete, 124

B

Bailey, Jerry L., 63
Baird, 2LT Jim, 122, 128-134
Ball, SP4 Gipsey, B., Jr., 31-34
Ballard, Ted, 33
Barrowcliff, Merton, 88
Bass, Sam, 124
Belcastro, SGT Dennis J., 35-37
Bell, SP4 David Leroy, 66
Bellino, COL Joseph O. "Joe," 26, 145
Bennett, Calvin, 129, 132-134
Bittner, SGT Roger, 124-126
Briggs, 1LT Thomas, 23, 24
Bledsoe, 22
Blink, Roger, 88
Bond, SGT Richard, 92, 94, 96
Bonner, Edward "Ed," 98
Boyd, Jim, 32
Brammer, SP4, 49
Branham, SP4 Vernon L., 69
Braswell, Baylon, 97
Britton, SSG Ronald, 71
Brockbank, LT Howard, 138
Buckwalter, SSG Jack, 115, 116
Buirge, Mike, 42, 44
Burke, LTC John, 27
Burnell, SGT John D., 63, 67, 68
Bush, Vernon, 40-44
Bushong, SP4, 49

C

Cahill, SGT Richard, 110, 111
Calhoun, J. D., 21, 22
Calibro, SP5 Steve, 61-66, 69
Callahan, 2LT Tom, 122, 133, 134
Callison, Jimmy Ray, 91, 92, 97
Cameric, Mike, 146
Cansans, PFC, 49
Capizola, SGT Nicholas, 58
Cappolloni, SP5 Dennis, 36, 37
Carter, 23
Carter, SP5 William Fred, 97, 98, 102, 103, 106
Chalker, Wayne, 117, 118

Champ, SGT Roger D., 69, 70, 73, 74
Chapman, SP5 John, 95, 99, 100, 102, 103
Christopher, Jerry, 31, 32
Cochran, 96
Cochran, Danny, 131
Cole, SGT Larry, 69
Collins, SSG Claude L., 24
Collins, SP4 Larry E., 113
Collins, SGT Leroy, 21, 24
Collins, MAJ Nicholas, 17
Colucci, CPT Ken, 127
Cox, SGT Buford, Jr., 22, 23
Crow, Walter, 124
Cummings, SP4 Harold, 35, 36
Czerwinsky, SP4, 32

D

Dahl, SP4 Larry, 92, 94, 96, 100-103
Daniels, LTC Walter C., 87
Darby, Jim, 111
Davison, SP4 Guy A., 141
Deeks, Walter, 95, 96, 99, 101-103, 106
Delmar, COL Henry "Hank," 114
Desena, 23
Diaz, SSG Hector J., 96, 100, 101, 103, 106
Dillahay, SP5 Robert, 37
Dodd, SGT John, 88-91, 143
Dominquez, SP4 Dick, 31
Dougherty, 1LT, 122
Downer, 131
Dye, SP4 Robert, 39-43

E

Edwards, SGT, 118
Ellis, LTC Alvin C. "Big Al," 133
Emery, 131

F

Fiandt, Larry, 79
Foster, Joseph, 35-37
Fowlke, SP4 Earnest W., 49
Frank, LT, 60
Frazier, SSG Charles H., 138, 141
Frazier, SP4 Richard B., 129, 132
Freeman, SP5 Erik, 91, 97, 100-103, 105
Fromm, Robert, 117
Fuller, 1LT Ralph, 121, 122, 133, 134

G

Giroux, Frank W., II, 35, 36
Geise, CPT Paul, 21, 23
Giap, Vo Nguyen, 34
Gordon, Ray E., 37
Green, SP4, 49

Greenage, PFC Roy L., 24
Gregory, SSG Kenneth R., 141
Gun Trucks
    54th Best, 63
    Ace of Spades, 63, 64, 66, 75, 77, 121, 134
    Assassins, 3, 125
    Baby Sitters, 124-127
    Bad Hombre, 103
    Ball of Confusion, 88, 91-93, 97
    Bel and the Cheese Eaters, 35
    Big Kahuna, 56
    Black Widow, 63, 69, 70, 77, 121, 129, 134
    Blood, Sweat and Tires, 81
    Boss, 89, 94-96, 99-103
    Brutus, 88-94, 96-98, 100-103, 106
    Cobra's Den, 111, 115
    Cold Sweat, 70, 71, 103, 141
    Corps Revenge, 62, 63, 67, 68
    Creeper, 94, 95, 98-100, 102, 106
    Ejaculator, 108, 111
    Eve of Destruction, 62-70, 73-75, 121, 123, 124
    Filthy Four, 63
    Flying Dutchman, 109, 111
    Gambler, 56, 63, 111, 115
    Grim Reaper, 63
    Grogin's Heroes, 111
    Hawk, 57
    Herm's Revenge, 68
    Iron Butterfly, 66, 67, 77-79
    Justifier, 125
    King Cobra, 81-83, 92
    King Kong, 63, 75, 77, 121, 122, 128-132
    Lady Be Good, 70, 71, 73
    Little Brutus, 96, 97, 100, 102
    Matchbox, 77-79
    Misfits, 88-91, 96, 100, 102, 103
    Mortician, 109, 111
    Outlaws, 91
    Peacemaker, 113
    Playboys, 94, 96, 99-103
    Poison Ivy, 81, 92
    Protector, 124, 125, 127
    Proud American, 132-134
    Roach Coach, 111
    Satan's Chariot, 104, 105
    Satan's Lil' Angel, 121, 128-132
    Sergeant Pepper II, 70
    Sir Charles, 81, 92, 100, 104, 105
    Steel's Wheels, 35, 37
    The Saint, 81, 83
    True Grit, 63, 69, 70, 75, 141
    Uncle Meat, 69, 70, 73, 75, 77, 79, 121, 128-132, 134, 146
    Untouchable, 91, 96-98, 100, 102, 103, 105, 106
    USA, 110, 111, 115
    War Wagon, 39, 44
    Widow Maker, 111
    Woom Doom, 91
Gunter, PFC William A., 24

H

Haley, SGT Gregory, 139
Han Jin Company, 58, 60, 65
Hardesty, SP4 Robert W., 53
Hastings, Ray "Corps," 63
Healy, SGT, 124
Heiser, MG Joseph M., Jr., 145
Hensinger, SP4 Arthur J., 31
Herman, Robert C. "Herm," 68
Himburg, LT James, 70-73
Hinote, SP4 William, 138, 139
Hish, Peter, 88-91
Hodges, John, 88-90
Honor, LTC Edward, 111, 112
Hoskins, LT Sam, 124
Howell, SP4, 49
Hughey, Lloyd R., 24
Huser, Charles L. "Chuck," 92, 94, 96, 100, 101, 103
Hunzeker, LTC William K., 23, 24
Hutcherson, SSG, 88, 90
Hutchins, LT, 67
Hyatt, PFC Gerald T., 37

I

Inuchi, Tony, 73
Israel, SGT Chester, 129, 130, 132, 143

J

Jackson, SP5 Jimmie, 49
Jacobs, John, 111
Johnson, PFC Bobby L., 141
Johnson, PFC Thomas G. "Tom," 69
Jones, Gene, 124, 126

K

Kagel, William "Bill," 91, 92
Kendal, SP4 Ron, 39, 41-45
Klepsig, CPL David E., 8
Kratzenmeier, SP4 Ivan, 139
Kunston, Glen, 97
Kurtz, 124

L

Larsen, LTG Stanley R., 26, 27
Larson, 131
Leclair, 23
Lee, SSG, 69, 70
Logan, Robert, 97, 146

Logston, Bob, 31, 32
Looney, Gary, 96, 99
Louden, 96, 99, 101
Loveall, 1LT Ronald C., 69
Ludy, COL Garland, 61-64, 69, 71, 76, 107, 117

M

Mack, 1LT Dennis C., 58
Maddox, Terrance N., 37
Mallory, Ronald "Ron," 92, 94, 96, 101, 103, 106
Maloney, 1LT Roger, 124-127
Manning, PFC Ronald, 83, 84
Marcum, Paul, 40-44
Mauricio, Paul, 95, 99-103, 106
McBride, SGT Michael "Mike," 125
McCarthy, CPT Joseph II, 69, 70, 73, 74
McDonald, 133
McEarchen, SGT Ellis "Mac," 94, 99
McGinley, LTC Edward, 61
McGinty, CPT Terrance J., 58
McGrath, James, 119
McGuire, Barry, 62
McQuellen, SGT, 101, 102
Melnick, 23
Merdutt, Gerald, 74
Metcalf, SP4 Charles E., 34
Midolo, 23
Mintz, SP4 Gary C., 66
Mongle, William, 94, 95
Montgomery, Barry, 94, 99, 103, 106
Morin, 22
Murphy, SP4 Albert, 138, 141

N

Newman, SP4 Bobby, 66

O

Ondic, Larry, 138
Orozco, Daniel, 139

P

Pacific & Atlantic Engineers, 48
Page, LTC John U. D. 7-11
Palumbo, 23
Parson, 70, 73
Patrick, CPT Wayne, 111-114, 116
Pedigo, SP4 Charles, 115
Pennington, 70
Phipps, 22
Plummer, Steve, 118
Porter, LT, 97, 98, 100, 102
Prescott, SGT Wayne, 88-90
Pulley, SP4 James E., 65, 75
Purvis, 1LT James P., 31

Q

Quintana, Ernest "Ernie," 91, 92

R

Reinhart, PFC Arthur W., 24
Republic of Korea Tiger Division, 16, 54, 70, 74
Revelak, LT, 126
Rippee, 71
Robertson, Walter, 111, 115
Ross, John, 97
Runkle, LTC Robert L., 37
Runnas, SP5 Stanley A., 37

S

Sanders, SP4 Roy A., 31-34
Sas, PFC Robert L., 31-34
Seay, SP4 William W., 138, 139, 141
Sellman, SP4 David M., 136, 138, 139, 141
Shartle, Harold, 96
Sheck, Kenneth, 96
Shed, Keith, 124
Shelley, 1LT Jay M., 65-67
Sherrill, Leroy, Jr., 125
Simmons, SP4 Ronald W., 24
Sims, SGT Charles, 104, 105
Smiley, LTC Philip N., 24
Smith, MG Oliver P., 7-9
Smith, Rick "Snuffy," 110
Smith, Ronald, 114, 115
Soule, SP4 Charles H., 124, 125, 127
Spears, 125
Spitler, CWO Robert J. "Bob," 139, 140
Spurgeon, SP4 Harold, 126
Stack, Robert, 97
Stafford, LT Lee, 124, 125
Stebner, PFC Robert L., 24
Stegmayer, 1LT Paul J., 53, 54
Steel, 1LT James R., 35, 37, 46
Swanson, COL Paul, 142
Sweeney, BG Arthur H. Jr., 127

T

Tate, Raymond H., 141
Taylor, Clifford W., 111
Thorne, SP4 Robert W. "Bob," 130, 132-134
Tidwell, PFC Jimmy, 49
Tillotson, 1LT, 37, 46
Tomlin, SGT, 23
Todd, 1LT Jerry L., 36, 37, 46
Trumbo, 23

U

US Army
    I Field Force, 26
    X Corps, 7, 8
US Army Divisions
    1st Cavalry Division, 22, 26, 30, 110, 117
    1st Infantry Division 140
    3rd Infantry Division, 7, 8
    4th Infantry Division, 26, 34, 48, 53, 73
    5th Infantry Division, 125, 133
    7th Infantry Division, 7, 10
    25th Infantry Division, 121, 136, 142
    101st Airborne Division, 112, 113, 117
US Army Regiments and Brigades
    1st Cavalry Regiment 33, 34, 48
    3rd Infantry Regiment, 134
    4th Cavalry Regiment, 139, 140
    5th Cavalry Regiment, 37
    5th Infantry Regiment, 37
    7th Cavalry Regiment, 22
    7th Infantry Regiment, 8
    17th Cavalry Regiment, 103
    22nd Infantry Regiment, 139, 140
    23rd Infantry Regiment, 137, 139, 140
    34th Armor Regiment, 137
    65th Infantry Regiment, 8, 9
    69th Armor Regiment, 55
    77th Armor Regiment, 125
    173rd Airborne Brigade, 40, 45, 48, 51-53
US Army Battalions
    1st Engineer Battalion, 9
    4th Battalion, 60th Artillery, 49, 52, 68
    25th Aviation Battalion, 139-141
    34th Supply and Service Battalion, 23
    65th Engineer Battalion, 139
    70th Engineer Battalion, 24
    240th Quartermaster Battalion, 38, 55, 88
    504th MP Battalion, 21, 22, 27
    720th MP Battalion, 136, 141
    815th Engineer Battalion, 27
    Task Force Faith, 7, 8
    Task Force Dog, 8-10
US Army Companies
    25th MP Company, 136, 138
    304th Supply and Service Company, 40, 44
    557th Maintenance Company, 111
    560th Maintenance Company (DS), 59
US Army Transportation Corps Groups
    8th Transportation Group, 15-19, 23-25, 27, 34, 38-40, 42-44, 53, 54, 56, 60-62, 68, 70, 71, 76, 81, 86, 97, 107, 111, 117, 122, 124, 127, 139, 144-148
    48th Transportation Group, 136, 138, 141, 142
    500th Transportation Group, 107, 108, 111, 115, 144-146
US Army Transportation Corps Battalions
    6th Transportation Battalion, 28, 136, 139 142, 145
    7th Transportation Battalion, 136
    24th Transportation Battalion, 107, 114
    27th Transportation Battalion, 16, 20, 24, 51, 52, 54, 55, 60, 70, 77, 81, 87, 91, 94, 118
    36th Transportation Battalion, 107, 110, 111, 114
    39th Transportation Battalion, 118, 120, 122, 128, 132, 133
    52nd Transportation Truck Battalion, 7-10
    54th Transportation Battalion, 16, 17, 20-22, 24, 30, 31, 35, 47, 48-52, 54, 55, 60, 67, 70, 77, 79, 104
    57th Transportation Battalion, 117, 120
    124th Transportation Battalion, 16, 27, 48, 49, 51, 53, 58, 88
US Army Transportation Corps Companies
    2nd Transportation Company, 53, 81
    57th Transportation Company, 24, 36, 117, 123-125, 127
    58th Transportation Company, 51
    61st Transportation Company, 117
    62nd Transportation Company, 137, 138, 141
    63rd Transportation Company, 117, 120
    64th Transportation Company, 1, 49, 53
    88th Transportation Company, 58-60, 84, 104
    172nd Transportation Company, 107
    205th Transportation Company, 74
    359th Transportation Company, 1, 23, 38-45, 88-103
    360th Transportation Company, 109, 111, 114, 115, 117
    363rd Transportation Company, 117
    377th Transportation Company, 7
    442nd Transportation Company, 107, 111, 112, 114
    444th Transportation Company, 51, 80, 81, 83
    446th Transportation Company, 117
    512th Transportation Company, 24, 31, 55, 65-67, 70, 77, 78
    515th Transportation Company, 123-125, 127
    523rd Transportation Company, 21, 24, 33, 55, 61-75, 77, 121-125, 127-134
    538th Transportation Company, 117
    545th Transportation Company, 94
    566th Transportation Company, 107, 110, 111, 115
    572nd Transportation Company, 117, 124
    585th Transportation Company, 30, 117, 118, 119, 123, 127
    597th Transportation Company, 81, 83, 85, 92, 100
    666th Transportation Company, 17, 21, 22, 30, 31
    669th Transportation Company, 24, 35-37, 66, 70
    670th Transportation Company, 107, 114
    863rd Transportation Company, 117
US Marine Corps
    1st Marine Division, 7, 8, 10
    1st Marine Regiment, 7, 8
    5th Marine Regiment, 8
    7th Marine Regiment, 8, 9
    1st Marine Regimental Train, 9
    7th Marine Motor Transport Battalion, 9
Usher, Jerry, 88

V

Van Brocklin, SGT Terrance, 70-73
Voightritter, CPT Donald "CPT V," 122, 133
Voightritter, CPT Ronald, 81-86

W

Winston, LTC Walden C., 8
Ward, Bill, 88-91
Wasson, PFC Marvin L., 10
Wayne, John, 69
Weeks, SGT Grailin, 96, 99, 102, 103
Wehunt, Billy D., 80
Welch, SGT, 49
Weston, Thomas, 34
Welton, 2LT Burrell, 23, 39-45
Wernstrum, Alan, 88-91
Werth, 146
Wheat, SP4 Timothy, 39, 41-44
Whitehead, Harrison "Bud," 74
Williamson, MG Ellis W., 137, 141
Wilson, Alan, 88
Wilson, 1LT David R., 49, 129
Wilson, PFC Earl C., 58
Witten, PFC Jackie W., 65
Wolf, COL Duquesne "Duke," 137, 139, 140
Wolfe, LTC Melvin M., 24, 27
Wood, SP4 Calvin, 82-86
Woodell, 22

Y

Young, 23